Exploring Medicinal Orchids

Orchids are fascinating ornamental flowers with a huge market both as cut flowers and potted plants, but they are also used in the traditional medicine system as they can be rich in phytochemicals with exceptional medicinal properties. This book, *Exploring Medicinal Orchids*, is a comprehensive guide for medicinally important orchids, their diversity, and their use in traditional and folk medicines. It presents information on secondary metabolites of medicinally active orchid species, pharmaceutical and medicinal applications, and describes advanced techniques of biotechnology in the conservation of medicinal orchids.

Features

- Dedicated chapters discuss phytochemical constituents and medicinal properties of orchids
- Enriches the understanding of secondary metabolites of medicinal orchids and their pharmaceutical application
- Elaborates the botanical, ethnomedicinal, and pharmacological aspects of various medicinal orchids
- Demonstrates tissue culture systems for the conservation of medicinal orchids

A volume in the *Exploring Medicinal Plants* series, this book is a resource for academics, researchers, professors, and students working in the fields of medicinal plants and flowers, biodiversity conservation, herbal medicine, and plant biotechnology.

Exploring Medicinal Plants

Series Editor: Azamal Husen, *Wolaita Sodo University, Ethiopia*

Medicinal plants render a rich source of bioactive compounds used in drug formulation and development; they play a key role in traditional or indigenous health systems. As the demand for herbal medicines increases worldwide, supply is declining as most of the harvest is derived from naturally growing vegetation. Considering global interests and covering several important aspects associated with medicinal plants, the Exploring Medicinal Plants series comprises volumes valuable to academia, practitioners, and researchers interested in medicinal plants. Topics provide information on a range of subjects including diversity, conservation, propagation, cultivation, physiology, molecular biology, growth response under extreme environment, handling, storage, bioactive compounds, secondary metabolites, extraction, therapeutics, mode of action, and healthcare practices.

Led by Azamal Husen, PhD, this series is directed to a broad range of researchers and professionals consisting of topical books exploring information related to medicinal plants. It includes edited volumes, references, and textbooks available for individual print and electronic purchases.

Promising Antiviral Herbal and Medicinal Plants
Edited by Nadeem Akhtar, Azamal Husen, Vagish Dwibedi, and Santosh Kumar Rath

Antimalarial Medicinal Plants
Edited by Azamal Husen

Medicinal Spice and Condiment Crops
Edited by Azamal Husen

Therapeutic Medicinal Plants in Traditional Persian Medicine
Edited by Roja Rahimi and Roodabeh Bahramsoltani

Medicinal Plants for the Management of Neurodegenerative Diseases
Edited by Jamal Akhtar, Maryam Sarwat, and Fouzia Bashir

Nutraceuticals Inspiring the Contemporary Therapy for Lifestyle Diseases
Edited by Mala Trivedi, Sachidanand Singh, Parul Johri, and Pedro López-Sánchez

Plants as Medicine and Aromatics: Uses of Botanicals
Edited by Mohd Kafeel Ahmad Ansari, Mushtaq Ahmad, and Gary Owens

Exploring Traditional Wild Edible Plants
Edited by Vibhor Agarwal, Sachidanand Singh, and Rhaul Datta

Propagation to Pharmacopeia: Modern Approaches in Medicinal Plants
Edited by Abha Manohar K, Gopal Shukla, Saikat Gantait, A P Das, and Sumit Chakravarty

Exploring Medicinal Orchids
Edited by Azamal Husen and Manu Pant

For more information about this series, please visit: www.routledge.com/Exploring-Medicinal-Plants/book-series/CRCEMP

Exploring Medicinal Orchids

Edited by
Azamal Husen and Manu Pant

CRC Press
Taylor & Francis Group
Boca Raton London New York

CRC Press is an imprint of the
Taylor & Francis Group, an **informa** business

Designed cover image: Manu Pant, Veronica Lang, T. Usha Bharathi, and Ashis Kumar Roy

First edition published 2025
by CRC Press
2385 NW Executive Center Drive, Suite 320, Boca Raton FL 33431

and by CRC Press
4 Park Square, Milton Park, Abingdon, Oxon, OX14 4RN

CRC Press is an imprint of Taylor & Francis Group, LLC

Library of Congress Cataloging-in-Publication Data
Names: Husen, Azamal, editor. | Pant, Manu, editor.
Title: Exploring medicinal orchids / edited by Azamal Husen and Manu Pant
Other titles: Exploring medicinal plants.
Description: First edition | Boca Raton, FL : CRC Press, 2025 | Series: Exploring medicinal plants | Includes bibliographical references and index | Summary: "Orchids are fascinating ornamental flowers having a huge market both as cut flowers and potted plants, but they are also used in the traditional medicine system as they can be rich in phytochemicals with exceptional medicinal properties. This book, Exploring Medicinal Orchids, is a comprehensive guide for medicinally important orchids, their diversity, and use in traditional and folk medicines. It presents information on secondary metabolites of medicinally active orchid species, pharmaceutical and medicinal applications, and describes advanced techniques of biotechnology in the conservation of medicinal orchids—Provided by publisher
Identifiers: LCCN 2024024279 (print) | LCCN 2024024280 (ebook) | ISBN 9781032555539 (hardback) | ISBN 9781032566528 (paperback) | ISBN 9781003436638 (ebook)
Subjects: LCSH: Orchids—Therapeutic use. | Medicinal plants.
Classification: LCC QK495.O64 E96 2025 (print) | LCC QK495.O64 (ebook)
LC record available at https://lccn.loc.gov/2024024279
LC ebook record available at https://lccn.loc.gov/2024024280

ISBN: 9781032555539 (hbk)
ISBN: 9781032566528 (pbk)
ISBN: 9781003436638 (ebk)

DOI: 10.1201/b23375

Typeset in Times Lt Std
by Apex CoVantage, LLC

Dedicated to the eternal learner Sh. Avadh Bihari Pant

(Father of Dr. Manu Pant)

Contents

Preface

Orchidaceae is the second largest family of flowering plants, including about 25,000 to 35,000 species that are widely cultivated for exceptionally beautiful flowers. The orchids are found worldwide and rank at the top in the global floriculture trade. Their therapeutic value is, however, generally underappreciated. In many regions of the world, the traditional and folk medical systems depend heavily on the local empirical knowledge of orchids. Numerous orchid species have been shown to contain bioactive substances like alkaloids, flavonoids, phenols, terpenoids, stilbenoids, phenanthrenes, etc. These substances are known to provide orchids with a plethora of medicinal characteristics, including antibacterial, antioxidant, anti-inflammatory, antiallergenic, aphrodisiac, and wound-healing properties. The effectiveness of orchids in treating conditions including convulsions, diarrhea, malaria, and bronchitis, as well as more serious illnesses like rheumatism, brain disorders, and cancer, has also been validated by individual studies. Despite these findings, detailed research on the medicinal benefits of orchids and their therapeutic characteristics is still lacking.

The International Union for Conservation of Nature's Red Data Book lists orchids, including medicinal orchids, as having declining natural populations. Irrational and unscientific harvesting of these valuable plants and plant parts from the wild has also led to their inclusion in Appendix II of the Convention on International Commerce in Endangered Species of Wild Fauna and Flora (CITES). Nevertheless, random collection from natural areas is still conducted to prepare traditional medicines and food or to meet the expanding needs of the pharmaceutical sector.

This book emphasizes the importance of thorough research on orchid species for potential medical applications. The book discusses in detail the diversity of medicinal orchids, their age-old uses in traditional medicine systems, secondary metabolites of medicinally active orchid species, and strategies for orchid conservation. Another concurrent matter of the subject of the book is addressed in chapters devoted to orchid species with proven medicinal characteristics and those that can be classified under a certain genus. The book will be very useful to readers ranging from graduate students, researchers, and growers interested in medicinally active orchid cultivation and sustainable use.

We extend our heartfelt gratitude to all the authors who enriched this book through their expert contributions. We also thank the publishers for supporting and facilitating us in the realization of this book. Their cooperation in all stages of book production is highly appreciated.

We believe that this book will contribute to an improved understanding of the subject and stimulate fresh ideas for sustainable utilization, research, and development in the field of medicinal orchids for a better tomorrow.

Azamal Husen and Manu Pant

About the Editors

Azamal Husen is Adjunct Professor at Graphic Era (Deemed to be University), Dehradun, Uttarakhand, India. He is also working as a Visiting Professor at the University Putra Malaysia, Selangor, Malaysia, and acting as a Foreign Delegate at Wolaita Sodo University, Wolaita, Ethiopia. Previously, he served as Professor and Head of the Department of Biology, University of Gondar, Ethiopia. He also worked as a Visiting Faculty of the Forest Research Institute and the Doon College of Agriculture and Forest at Dehra Dun, India. His 25 years of research and teaching experience encompasses biogenic nanomaterial fabrication and application; plant responses to nanomaterials; plant adaptation to harsh environments at the physiological, biochemical, and molecular levels; herbal medicine; and clonal propagation for improvement of tree species. He has conducted research sponsored by the World Bank, the National Agricultural Technology Project, the Indian Council of Agriculture Research, the Indian Council of Forest Research Education, and the Japan Bank for International Cooperation. Husen has published extensively (~300) and served on the editorial boards and as a reviewer of reputed journals published by Elsevier, Frontiers Media, Taylor & Francis, Springer Nature, RSC, Oxford University Press, Sciendo, the Royal Society, CSIRO, PLOS, MDPI, John Wiley & Sons, and UPM Journals. He is also on the advisory board of Cambridge Scholars Publishing, UK. He is a fellow of the Plantae group of the American Society of Plant Biologists and a member of the International Society of Root Research, Asian Council of Science Editors, and International Natural Product Sciences. He is Editor-in-Chief of the *American Journal of Plant Physiology* and a Series Editor of *Exploring Medicinal Plants* (Taylor & Francis Group, USA); *Plant Biology, Sustainability, and Climate Change* (Elsevier, USA); and *Smart Nanomaterials Technology* (Springer Nature, Singapore). He has achieved the distinguished honour of being recognized as one of the "World's Top 2% Scientists" for the year 2022, 2023 and again for 2024 by Stanford University, USA. This recognition has also been prominently featured in the Elsevier Data Repository.

Manu Pant is currently an Associate Professor and Head of the Department of Biotechnology at Graphic Era Deemed to be University, Dehradun, Uttarakhand, India. She received her master's degree in Applied Microbiology and Biotechnology from Banasthali Vidyapith in Rajasthan, India, and her doctorate in Biotechnology from HNB Garhwal University (A Central University) in Srinagar, India. She has extensive experience in the field of plant tissue culture, having worked as a research fellow at the Forest Research Institute and the Indian Council of Forestry Research and Education on nationally funded projects. Her research work on orchids has been funded by the state governments, earning her recognition and awards at the national level. She has several peer-reviewed international papers to her credit, published patents, and delivered numerous oral and poster presentations, keynote addresses, and invited talks at national and international scientific events. Pant has conducted many national and international conferences, seminars, and trainings related to the thrust areas in plant sciences. She has been supervising PhD, MSc, and BSc students in their research work on different aspects of orchids, medicinal plants, and indigenous crop species. For her work in the field of plant sciences, Pant has received appreciation from State and National level bodies.

Contributors

Gopalakrishnan Aazhivaendhan
Department of Botany
Bharathidasan University
Tiruchirappalli, Tamil Nadu, India

Himani Adhikari
Department of Biotechnology
Graphic Era Deemed to be University
Dehradun, Uttarakhand, India

Manuj Awasthi
Department of Horticulture
Chandra Shekhar Azad University of Agriculture and Technology
Kanpur, Uttar Pradesh, India

T. Usha Bharathi
Division of Flower and Medicinal Crops
Indian Institute of Horticultural Research
ICAR-Indian Institute of Horticultural Research
Bengaluru, Karnataka, India

V. Bhargav
Multi-technology Testing Centre and Vocational Training Center
College of Agriculture
Pasighat, Arunanchal Pradesh, India

Kiran Bisht
Department of Biotechnology
Graphic Era Deemed to be University
Dehradun, Uttarakhand, India

Abhishek Chandra
Department of Floriculture and Landscape Architecture
Sardar Vallabhbhai Patel University of Agriculture and Technology
Meerut, Uttar Pradesh, India

Nital Nilesh Chankeshwara
Department of Biotechnology
Graphic Era Deemed to be University
Dehradun, Uttarakhand, India

Kritika Chauhan
Department of Biotechnology
Graphic Era Deemed to be University
Dehradun, Uttarakhand, India

Ashish Gaur
Department of Biotechnology
Graphic Era Deemed to be University
Dehradun, Uttarakhand, India

Aarushi Gautam
Department of Biotechnology
Graphic Era Deemed to be University
Dehradun, Uttarakhand, India

S.M. Khasim
Department of Botany and Microbiology
Acharya Nagarjuna University
Guntur, Andhra Pradesh, India

Thiruppathi Senthil Kumar
Department of Botany
Bharathidasan University
Tiruchirappalli, Tamil Nadu, India

Anamika Kumari
Department of Botany
Goswami Ganesh Dutta Sanatan Dharma College
Chandigarh, Punjab, India

Suman Kumaria
Plant Biotechnology Laboratory
Department of Botany
North-Eastern Hill University
Shillong, Meghalaya, India

Vikas Kushwaha
Department of Biotechnology
Goswami Ganesh Dutta Sanatan Dharma College
Chandigarh, Punjab, India

Sunil Malik
Department of Floriculture and Landscape Architecture

Sardar Vallabhbhai Patel University of Agriculture and Technology
Meerut, Uttar Pradesh, India

Kalkame Ch. Momin
Multi-technology Testing Centre and Vocational Training Center
College of Agriculture
Central Agricultural University
Imphal, Manipur, India

S. Pooja Murthy
Division of Flower and Medicinal Crops
Indian Institute of Horticultural Research
ICAR-Indian Institute of Horticultural Research
Bengaluru, Karnataka, India

Mariappan Muthukumar
Department of Botany
Bharathidasan University
Tiruchirappalli, Tamil Nadu, India

Ayushi Negi
Department of Biotechnology
Graphic Era Deemed to be University
Dehradun, Uttarakhand, India

Promila Pathak
Department of Botany
Panjab University
Chandigarh, Punjab, India

Mukti Ram Paudel
Central Department of Botany
Tribhuvan University
Kirtipur, Nepal

Prasenjit Paul
Department of Botany
St. Edmund's College
Shillong, Meghalaya, India

Mukti Ram Poudeyal
Bhairahawa Multiple Campus
Tribhuvan University
Rupandehi, Nepal

Anu Prabha
Department of Botany
Goswami Ganesh Dutta Sanatan Dharma College
Chandigarh, Punjab, India

M. Rahamtulla
Department of Botany and Microbiology
Acharya Nagarjuna University
Guntur, Andhra Pradesh, India

Chinnaiyan Rajasekar
Department of Botany
Alagappa University
Karaikudi, Tamil Nadu, India

Venkatesh Rampilla
Department of Botany
Government College (Autonomous), Rajamahendravaram
Rajahmundry, Andhra Pradesh, India

J. Ramudu
Department of Botany
Dr. L.H.R. Government Degree College
Mylavaram, Andhra Pradesh, India

Pawan Singh Rana
Plant Tissue Culture Lab, Netala
Department of Horticulture and Food Processing
Uttarkashi, Uttarakhand, India

Abhijeet Rathod
Department of Biotechnology
Graphic Era Deemed to be University
Dehradun, Uttarakhand, India

Khageshwari Saud
Central Department of Botany
Tribhuvan University
Kirtipur, Nepal

B.D. Sharma
Distinguished Professor and Curator Herbal Garden
Eternal University
Baru Sahib, Sirmour, India

Uday Kumar Singh
Pokhara Campus
Tribhuvan University
Pokhara, Nepal

Govindarajan Siva
Department of Botany
Bharathidasan University
Tiruchirappalli, Tamil Nadu, India

Noyaljohnson Tinoammini
Department of Botany
Bharathidasan University
Tiruchirappalli, Tamil Nadu, India

T. Yatung
Multi-technology Testing Centre and Vocational Training Center
College of Agriculture
Imphal, Manipur, India

1 Medicinal Orchids

Potential, Conservation, and Healing Traditions

Ayushi Negi, Kritika Chauhan, Himani Adhikari, Azamal Husen, and Manu Pant

1.1 INTRODUCTION

Orchids are unquestionably the most intriguing and diversified collection within the domain of monocots in the enormous angiosperms, as over 25,000 species have been grouped into over 750 genera that are known to exist. The vibrant flowers of orchids, which are valued for their aesthetic appeal, are frequently used in floral compositions, decorative arrangements, and horticultural endeavors and are the primary motivation for orchid cultivation (Panda & Mandal, 2013). India is a great reservoir for these attractive plants due to its diversity in orchid's genetic resources. An estimated 1,300 different orchid species are thought to be present within the political territories of India, exhibiting incredible and unmatched varieties. Orchids have been used in different parts of the world as an ancient healing practice as well as the treatment of various diseases (Lal et al., 2020). The historical usage of orchids in India can be traced back to the Vedic period which dates back to 2500 BCE, and the Ayurvedic school of traditional medicine, where their significance was duly documented. Ayurvedic, Siddha, Unani, and traditional tribal systems of medicine have all found a substantial amount of use for close to 150 of these diverse species. Indian orchids' therapeutic value in treating conditions like hypertension and allergies, mental disorders, burns, skin issues, tuberculosis, rheumatism, and malignancy are all well-known in the existing literature. Some of the orchid species used for medicinal purposes from ancient times include *Aerides odorata, Geodorum densiflorum*, and *Epipactis royleana* (Figure 1.1(a,b,c)), and many others whose therapeutic benefits have been recognized and incorporated in age-old healing techniques (Vaidya, 2019; Pal et al., 2022).

Although orchids are predominantly cultivated for ornamental purposes, they also have nutritional as well as cultural values for different tribes all around the world. Indigenous communities residing in the Northeastern region have long incorporated wild orchid species into their dietary practices. The leaves of *Cymbidium* species, in particular, have been used as a source of nutrition by the local tribes. They produce a savory sauce to complement their cereal-based meals by crushing the fresh shoots of this species (De et al., 2014). This ancient culinary application highlights the tremendous relationship these people have with their natural surroundings, where the abundant gifts of wild orchids are loved and utilized in a way that feeds both body and culture. Many medicinal orchid species, including *Habenaria acuminata, Habenaria susannae, Cephalanthera ensifolia* (Figure 1.2), *Microstylis wallachii, Orchis latifolia*, and *Satyrium* hold significant importance in a nutritious diet. These orchids' pseudobulbs, roots, and rhizomes are used to supply critical nutritional demands and contribute to the socioeconomic well-being of the populations in this area (Medhi & Chakrabarti, 2009). It is important to note that while the utilization of orchids for decorative purposes is on a high, there has been very less attention paid very less attention has been paid to their medicinal qualities and harnessing their potential for therapeutic benefits.

DOI: 10.1201/b23375-1

FIGURE 1.1 a) *Aerides odorata*, b) *Geodorum densiflorum*, c) *Epipactis royleana.*

FIGURE 1.2 *Cephalanthera ensifolia.*

1.2 HISTORICAL PERSPECTIVE

The Chinese are credited with being the first to produce and describe orchids, as well as the ones to characterize them as having medical properties. According to ancient Chinese records, *Dendrobium* is a source of astringent, analgesic, and anti-inflammatory chemicals. The word "Vanda" is used to refer to plants in the Indian Vedic literature. Surprisingly, one of the most visually attractive groups of orchids now includes a genus with the same name (Jalal et al., 2008). The preparation techniques for formulations, including orchids combined with other plants and their therapeutic properties, are described in ancient texts, such as the Rig-Veda. Many orchids play a significant role in conventional medical systems because they are rich in cardiac glycosides, flavonoids, glycosides, phlobatanins, quercetin, reducing sugars, saponins, steroids, alkaloids, anthraquinones, terpenoids whose bioactivity has been determined (Vaidya, 2019). In Ayurveda "Astavarga" is essential in the creation of several tonics, such as the well-known "Chyawanprash," as well as rejuvenating substances. "Astavarga" includes eight different botanical species out of which four orchid species held particular importance: *Malaxis acuminata* D. Don (Rishbhaka), *Malaxis muscifera* (Lindl.) Kuntze (Jeevak), *Habenaria edgeworthii* Hook. f. ex Collett (Vriddhi), and *Habenaria intermedia* D. Don (Riddhi). These famous tonics and rejuvenation mixtures were made using these orchids, which were prized for their special therapeutic capabilities. The persistent use of orchids in these antiquated procedures highlights their ongoing significance and contribution to India's traditional medical practices, preserving its cultural and botanical legacy over the decades (Kant et al., 2012).

1.3 CHEMICAL COMPOSITION OF ORCHIDS

Medicinal orchids are extremely rich in alkaloid content. Experimental evidence highlights the isolation of various alkaloids, including anthocyanins, stilbenoids, and triterpenoids, from orchids. Notable compounds such as orchinol, hircinol, cypripedin, jibantine, nidemin, and loroglossin have been reported from orchids (De & Pathak, 2015). The family Orchidaceae is well recognized for generating volatile organic compounds (VOCs), which serve as an indicator for pollinators and are found all over the world. However, there is a paucity of studies on the volatile profiles of orchids. In a study conducted in Italy (Robustelli della Cuna et al., 2022), GC/FID (Gas Chromatography With Flame Ionization Detection) and GC/MS (Gas Chromatography-Mass Spectrometry) were used to examine essential oils from sympatric orchid species. The detected substances included alcohols (16.93–50.60%), with p-cresol (12.75–38.10%) standing out, saturated hydrocarbons (5.81–59.29%), with pentacosane (2.22–40.17%) and tricosane (0.78–27.48%) being notable, and long-chain monounsaturated hydrocarbons (0.29–5.20%). In *Orchis purpurea*, coumarin predominated (68.84%); it was found in lesser levels (0.21–0.26%) in other species. Specific substances might only be present in specific orchid species, but they undoubtedly play a significant role in pollinator interactions.

A study by Natta et al. (2022) highlighted six endangered orchid species *Dendrobium nobile*, *D. moschatum* (Figure 1.3), *D. densiflorum*, *Acampe papillosa*, *Coelogyne nitida*, and *Arundina graminifolia* in the northeastern Himalayan regions for their potential medicinal properties. The study confirmed the orchids' phytochemical composition and antioxidant properties using spectrophotometric, AAS (Atomic absorption spectrometry), HPLC (High Performance Liquid Chromatography), and GCMS (Gas chromatography/Mass spectrometry) methods. While *Acampe papillosa* displayed considerable phenols, tannins, and saponins, *D. moschatum* showed large levels of carbohydrates and protein, *D. densiflorum* had rich flavonoids and alkaloids with strong ABTS scavenging, and *Coelogyne nitida* had intense DPPH scavenging, and mineral levels were higher in *Dendrobium* species. Table 1.1 lists some orchid species with the associated phytochemical compounds.

1.4 ORCHIDS IN HEALTHCARE

Many studies have shown that the different parts of orchids, like tubers, rhizomes, leaves, roots, and pseudobulbs, are utilized in various ways, such as their paste, extract, and powder, for the treatment of a wide range of diseases and disorders. *Acampe, Aerides, Bulbophyllum, Cephalanthera, Epipactis, Flickingeria, Gastrodia, Goodyrea, Listera, Satyrium,* and *Vanda* are some of the highly utilized medicinal orchid genera. Orchid species such as *Aerides multiflora, Malaxis acuminata, Bulbophyllum morphologlorum, Spiranthes australis,* etc., also exhibit anti-microbial activities.

A study by Sandrasagaran et al. (2014) illustrated the potential anti-microbial activity of *Dendrobium crumenatium* using different parts of the plant. The study was conducted against eight pathogenic bacteria using a disc diffusion assay, microdilution test for determination of minimum inhibitory concentration (MIC) and minimum bacterial concentration (MBC). Through their study, it was found that the methanolic extract of *Dendrobium crumenatium* stem had the most considerable antimicrobial activity against *Staphylococcus aureus, Klebsiella pneumoniae,* and *Enterobacter aerogenes*. Also, its root and stem extract was found to be effective against *Streptococcus pneumoniae,*

FIGURE 1.3 *Dendrobium moschatum.*

TABLE 1.1
Important Phytochemical Compounds Isolated from Orchids

S. No.	Name of Species	Phytochemical Compounds	References
1.	*Anacamptis coriophora*	• Methyl-(E)-p-methoxycinnamate • 13-heptadecyn-1-ol,7,9-di-tert-butyl-1-oxaspiro (4,5) deca-6,9-diene-2,8-dione • 10-dodecenol, p-cresol • methyl (Z) p-methoxycinnamate • 2-dodecenal, methyl cinnamate • 2,5-dimethoxybenzyl alcohol • 4-(1,1,1,3,3-tetramethylbutyl)-phenol	Mokni et al., 2016
2.	*Amacamptis pyramidalis*	• The high content of flavonoids, chlorophyll a-b and reduced glutathione	Štajner et al., 2010

(Continued)

TABLE 1.1 ***(Continued)***
Important Phytochemical Compounds Isolated from Orchids

S. No.	Name of Species	Phytochemical Compounds	References
3.	*Cypripedium calceolus*	• Linalool • octyl acetate • 18 aromatic compounds • 32 terpenoid derivatives • 22 aliphatic	Braunschmid et al., 2020
4.	*Dactylorhiza maculata*	• 4hydroxyphenylacetic acid methyl ester, • 4-hydroxybenzaldehyde, • 4-hydroxybenzaldehyde alcohol	Naczk et al., 2018
5.	*Dactylorhiza majalis*	• Lactic acid • nonanoic acid • caproic acid • palmitic acid • p-hydroxybenzyl alcohol	Wróblewska et al., 2019
6.	*Dactylorhiza incarnata*	• 4-hydroxyphenylacetic acid methyl ester • 4-hydroxybenzaldehyde • 4-hydroxybenzyl alcohol	Naczk et al., 2018
7.	*Dactylorhiza osmanica*	• Flavonoids • fumaric acid • p-coumaric acid • quercitrin • vanillic acid • ascorbic acid	Kiziltas et al., 2022
8.	*Goodyera repens*	• Kinsenoside • Gastrodin • Goodyeroside A	Wu et al., 2020
9.	*Himantoglossum caprinum*	• Glucomannan	Acemi et al., 2019
10.	*Platanthera bifolia*	• Benzyl benzoate • benzyl salicylate • methyl salicylate • cinnamyl alcohol • lilac aldehydes • methyl benzoate • Hexadecenoic acid methyl ester	D'Auria et al., 2020
11.	*Pseudorchis albida*	• Limonene, β-myrcene • (Z)-verbenol, verbenone • 4-oxoisophorone • β-phellandrene • (E)-sabinene hydrate • (Z)-sabinene hydrate • β-sabinene	Jersáková et al., 2011
12.	*Serapias vomeracea*	• Glucomannan	Acemi et al., 2019

Shigella dysentriae, and *Saccharomyces cerevisiae,* which was more effective than the standard antibiotics such as amoxicillin, chloramphenicol, and kenamycin. Many other orchid species have such potential medicinal properties that could be due to the presence of high alkaloid and flavonoid compounds. Table 1.2 gives a comprehensive view of a variety of medicinal uses of orchids.

TABLE 1.2
Some Common Uses of Medicinal Orchids

S. No.	Name of Species	Parts Used	Medicinal Uses	References
1.	*Aerides multiflora*	Leaf	Treating wounds, antibacterial properties, anti-diabetic potential	Lal et al. 2023
2.	*Acampe carinata*	Root	Used to treat stomach problems and chest pain that happens due to excessive acidity. Root paste was applied superficially on the bites of scorpions and snakes	Dash et al., 2008
3.	*Acampe papillosa*	Root	Used as the remedy to cure rheumatism and uterine diseases	De & Pathak, 2015
4.	*Acampe praemorsa*	Root	Used in the treatment of rheumatism	Dash et al., 2008
5.	*Aerides multiflora*	Tuber	Antibacterial property	De & Pathak, 2015
6.	*Aerides odorata*	Root, leaves	Treatment of joint pain, swelling, wounds, and tuberculosis	Dash et al., 2008
7.	*Anoectochilus roxburghii*	All parts	treatment of cancer, hypertension, lung disease, renal insufficiency, and other incurable illnesses, general tonics. Clinical evidences for effectiveness in treating *Helicobacter pylori* infection, type 2 diabetes mellitus, chronic hepatitis B, cough-variant asthma, hyperuricemia, and hand, foot, and mouth illness	Gam et al., 2020
8.	*Bletilla striata*	Tuber	Frequently used to heal hematemesis, pulmonary hemorrhage, gastrorrhagia, and traumatic bleeding.	He et al., 2017
9.	*Bulbophyllum morphologlorum*	Leaf, pseudobulb, root	Used to cure chronic inflammation, lower fever, and treat tuberculosis	Huehne et al., 2020
10.	*Cephalanthera longifolia*	Rhizome	Anti-debilitating property	Shapoo et al., 2013
11.	*Cymbidium aloifolium*	Root, entire plant	Used in the treatment of paralysis, weakness of eyes, and vertigo	Dash et al., 2008; De & Pathak, 2015
12.	*Dactylorhiza hatagirea*	Tubers, flowers	Treatment of cough and cold, headache, fracture, stomachache, diarrhea, wound healing, effective in curing general weakness after delivery	Shapoo et al., 2013
13.	*Dendrobium crumenatum*	Stem	Antimicrobial activity	Sandrasagaran et al., 2014
14.	*Dendrobium crysanthum*	Leaves	Antipyretic agent	Sandrasagaran et al., 2014
15.	*Dendrobium fimbriatum*	Leave	The paste of leaves is used over fractured areas to set the bones; it is used to boost body fluids	Sandrasagaran et al., 2014
16.	*Dendrobium moschatum*	Leaves	Treatement of earache and general body weakness	Medhi & Chakrabarti, 2009
17.	*Epipactis helleborine*	Rhizome, leaves, tubers	Used for fever, boils, and also consumed as a nerve tonic and aphrodisiac	Shapoo et al., 2013
18.	*Epipactis royleana*	Rhizome, tuber	Used to heal wounds, seminal debility, and general weakness	Shapoo et al., 2013

(Continued)

TABLE 1.2 ***(Continued)***
Some Common Uses of Medicinal Orchids

S. No.	Name of Species	Parts Used	Medicinal Uses	References
19.	*Eria bambusifolia*	Entire plant	Treatment of stomach ailments and hyperacidity	Dash et al., 2008
20.	*Flickingeria macraei*	Root	Effective in treating dermatitis and is also applied on affected parts to cure eczema	Dash et al., 2008
21.	*Gastrodia elata*	Rhizome	Treatment of treat seizure disorders, migraine, infantile spasms, tetanus, limb numbness, and drowsiness	Zhan et al., 2016; Su et al., 2023
22.	*Geodorum densiflorum*	Root	Commonly used to regularize the menstrual cycle	Dash et al., 2008
23.	*Goodyrea repens*	All parts	Treatment of toothache, healing wounds, loss of appetite, urinary irritation, irregular mensuration, and insect bites	Shapoo et al., 2013
24.	*Habenaria marginata*	Tuber	Used to cure malignant ulcers	Dash et al., 2008
25.	*Liparis odorata*	Leaves	Leaves extract used in burns, gangrene, and cancerous ulcers	Medhi & Chakrabarti, 2009
26.	*Listera ovata*	Rhizome, entire herb	Skin diseases, indigestion, and useful in stomach ailments	Shapoo et al., 2013
27.	*Luisia trichorhiza*	Root	Root extract is used as anti-diarrhea (in cattle) and to reduce myalgia in humans	Dash et al., 2008
28.	*Orchis latifolia*	Tubers, roots	Used as a nerve tonic and aphrodisiac	Medhi & Chakrabarti, 2009
29.	*Polystachya concreta*	Tuber	Used for the treatment of arthritis	Dash et al., 2008
30.	*Rhyncostylis retusa*	Roots, Leaf	Antioxidant, anti-diabetic properties, treatment of asthma, tuberculosis, menstrual disorders, rheumatism	Lal et al., 2023
31.	*Satyrium nepalense*	Pseudobulb	Used as a tonic, and it is used to cure dysentery and malaria	Medhi & Chakrabarti, 2009
32.	*Spiranthes australis*	Root	Treatment of Diabetes mellitus and urinary issues. Treatment of cancer, blood, and chest illnesses, as well as bacterial and inflammatory diseases	Peng et al., 2008
33.	*Spiranthes sinensis*	Flower, tubers	Skin eruption, swelling, sore throat, and in weakness	Shapoo et al., 2013
34.	*Vanda coerulea*	Leaves	Treatment of eye diseases and extract of leaves is used to treat diarrhea and external skin diseases	Medhi & Chakrabarti, 2009
35.	*Vanda tessellata*	Root	Root paste is used to heal rheumatism and neurotic disorders	Dash et al., 2008
36.	*Rhynchostylis retusa*	Roots	Roots are efficient against kidney stones, menstrual disorders, vertigo, asthma, and tuberculosis	De & Pathak, 2015
37.	*Pholidota chinensis*	Pseudobulbs	Healing toothache and stomachache	De & Pathak, 2015
38.	*Malaxis acuminata*	Pseudobulbs	Tonic of pseudobulbs is used to treat the burning sensation, tuberculosis, fever, and also used for enhancing sperm production	De & Pathak, 2015

(Continued)

TABLE 1.2 *(Continued)*
Some Common Uses of Medicinal Orchids

S. No.	Name of Species	Parts Used	Medicinal Uses	References
39.	*Habenaria intermedia*	Leaves, roots	Used to cure blood-related diseases	De & Pathak, 2015
40.	*Habenaria pectinata*	Leaves, tubers	Used in the treatment of arthritis	De & Pathak, 2015
41.	*Habenaria repens*	Tubers	Used as an aphrodisiac	De & Pathak, 2015

1.5 ALTERNATIVE PRODUCTS OF ORCHIDS

Other than medicinal uses, many orchids are also used to make value-added products that have high demand in the market. Some of these products and additional uses of orchids are mentioned in what follows.

1.5.1 Essential Oils

Orchids have an extensive range of floral smells. Nearly all natural scents may be produced by orchids, and their olfactory profiles exhibit a surprising complexity (Kaiser, 1993). Many orchids contain a wide variety of bioactive chemicals, including vitamins, antioxidants, hydrocolloids, proteins, etc. The skin, hair, and other body parts can be soothed, cleaned, restored, healed, and protected because of their beneficial qualities. Orchids show antioxidant activities that control the protective effects of oxidative stress on dermal cells. Reduced skin melanin pigment production and fostering a more even and healthy complexion production is one such aspect (Sharma & Pathak, 2020). The volatile chemicals that account for orchid fragrance are present inside plant structures. These molecules are stored as essential oils, which are kept in specialised cells called osmopheres that are found on the surface of flowers, leaves, or roots. Orchids only keep trace amounts of these substances since they can be hazardous to the plant. Volatile substances make up these aromatic oils and one smell them because they quickly vaporize at normal temperatures (De & Pathak, 2015). The enormous diversity of orchid species and the development of pollination techniques are both responsible for the wide range of floral smells found in orchids (Pridgeon et al., 2001; Pellegrino et al., 2008). Some chemical studies (Bhatnagar et al.,1971; and Gupta et al., 1978 revealed that the essential oil of *Crepidium acuminatum* contains limonene, eugenol, citronellal, 1–8-cineole, pipertone, and p-cymene. These substances include one sterol, β-sitosterol, a type of alcohol called ceryl alcohol, two different sugars called glucose and rhamnose, and five basic substances, one of which is choline. The pseudobulb of *Crepidium acuminatum* is renowned for its ability to treat several medical conditions, including bleeding diathesis, burning sensation, dipsia, and fever. It is also known to have a cooling effect, febrifuge properties, anti-aging properties, aphrodisiac qualities, and refrigerant properties (Adams et al., 2023). One of the most well-known products made from orchids is vanilla, an aromatic essence produced from the seed pods of the Vanilla orchid. The most significant vanilla species is *Vanilla planifolia*, which the Spanish introduced to Europe in 1510 (Berliocchi et al., 2004). In modern times, medical applications of *vanilla* are restricted to reducing nausea and increasing food ingestion in chemotherapy patients. Given that the early stages of Alzheimer's disease sometimes appear through a loss of the sense of smell, it is also used as a diagnostic scent for the illness (Lynn et al., 1992; Fladby et al., 2004). As a flavoring ingredient in baking, dried vanilla seed pods—particularly those from *Vanilla planifolia*—have substantial commercial value. They are also commonly used to add their distinctive and pleasant aroma in the production of perfumes and aromatherapy (Lubinsky et al., 2008). Essential oils are also used as the main therapeutic agents in aromatherapy, a supplementary therapy, to treat different types of medical ailments. These volatile or essential oils are extracted using several techniques from various plant parts, including barks, stems, flowers, roots, and other plant components (Babar et al., 2015). The essential oils from orchids are extremely popular in the aromatherapy community (Bose et al., 1999),

1.5.1.1 Salep

The Turkish word "Salep" is used to describe the tubers that come from terrestrial orchids. This word was used during the Ottoman era to refer to both the beverage made from these tubers and the orchid bulbs used in its preparation. The dried tubers of orchid species like *Orchis mascula, Orchis militaris*, and others are ground into salep, which is essentially a flour. These tubers are abundant in glucomannan, a healthy polysaccharide that resembles starch (Hossain, 2011). Due to the glucomannan in salep, it is used to bind ingredients while making ice cream. Additionally, it slows down the rate at which ice cream melts (Sezik, 2002). Salep has been highly valued as a dietary ingredient for patients with chronic diarrhea, extended physical weakness, and muscular and body pains (Friesen & Friesen, 2012). It has also been proven to be beneficial to overcoming indifference towards work and sleep as well as for treating a resistance to prolonged physical exercise.

1.5.1.2 Perfumery

Approximately 75% of all orchids are considered to be "fragrant," emitting discernible chemical compounds. These emissions can range from highly pleasant and fragrant odors to intensely repulsive smells in certain cases. Due to their complexity and compatibility with the physiological and chemical properties of the target pollinator, only certain odoriferous compounds are detectable by the human olfactory system (Salzmann et al., 2007). The *Catasetinae* and *Gongorinae* subtribes of orchids are known for producing the highest amounts of smell. The Lady of the Night orchid *(Brassavola nodosa*) releases a medicinal sweet odor after sunset, which reaches its higher intensity around midnight and fades away after daylight. In the morning, *Clowesia rosea* (Figure 1.4) emits a perfume resembling Vicks VapoRub, which changes to a cinnamon-like aroma in the afternoon. Another orchid, *Catasetum expansum,* emits a turpentine-like perfume in the morning and a rye bread-like scent as the afternoon wears on (De & Pathak, 2015).

1.6 MARKET VALUE OF ORCHIDS WORLDWIDE

The escalating market demand has established a strong orchid market at the global level. The dual value of ornamental and medicinal plants has further strengthened the economic significance of orchids.

The Netherlands is the top exporter of orchids, with a substantial 39.67% market share. Taiwan, Singapore, and New Zealand each contribute 10% of the market, closely trailed by Thailand with 28.41%. The final 6% of the export market is made up of various other nations (Figure 1.5b). Japan accounts for 30% of all imports worldwide, making it the top importer. With percentages of 12%, 10%, 7%, and 7%, respectively, the United Kingdom, Italy, France, and the United States are also significant importers (Figure 1.5a) (De & Medhi, 2015). With increasing reliance on herbal medicines, the demand for orchid-based therapeutic drugs is expected to increase significantly (De, 2023). This would create an improved market for therapeutic orchids that have been used in the traditional systems of medicine across the world (Behera et al., 2013). Given that almost 80% of the world's population receives medical treatment from herbal medicines, it is suggested that with their profusion of bioactive phytochemicals, have great potential as a source of therapeutic substances.

1.7 CONSERVATION AND SUSTAINABLE UTILIZATION

Considering the medicinal value of orchids, a methodical and scientific approach must be designed to ensure their judicious utilization (Mudoi et al., 2023). It is also important to understand that several

FIGURE 1.4 *Clowesia rosea.*

orchids that were present in abundance in their natural environments have declined in many parts of the world, majorly due to anthropogenic activities, unscientific collection, lack of awareness, loss of habitat, climate change, etc. Irrational harvesting practices not only limit the orchids' ability to recover organically, but they also have a long-lasting impact on their native environments (Ticktin et al., 2023). Consequently, orchids now occupy a prominent position in the IUCN Red Data Book (Fay, 2018). In 2017 the entire orchidaceae family was included in CITES' Appendix II, which lists prohibited international wildlife trade. Subsequently, the trade of orchids is strictly regulated internationally to avoid unsustainable exploitation and save these plants from additional damage brought

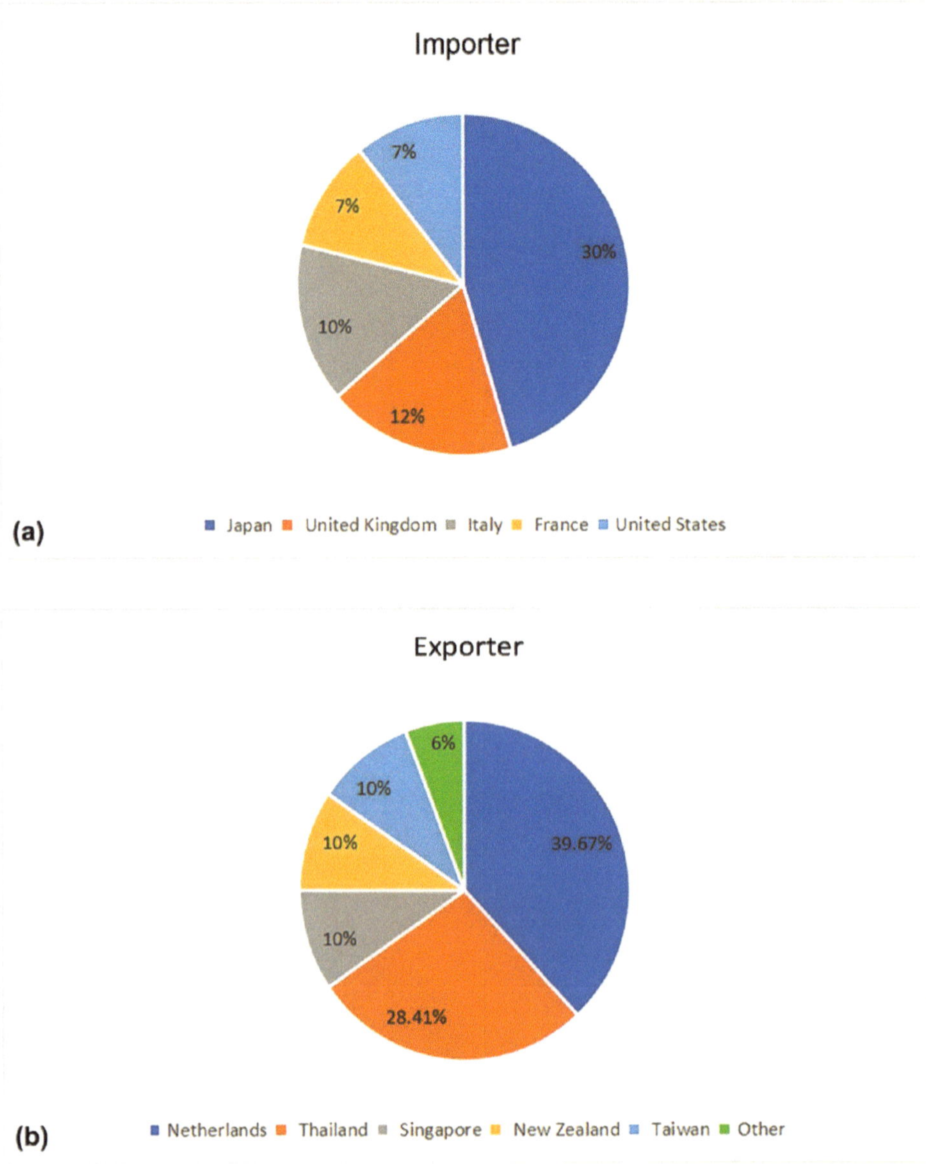

FIGURE 1.5 a) Major countries involved in export of orchids. b) Major countries involved in import of orchids.

on by excessive collection and trade (Pant, 2013). Despite these efforts, there is a major gap in the demand and supply of orchids. Conventional methods of propagation through seeds or keikis are challenged with limitations of high habitat specificity, reliance on pollinators, requirement for mycorrhiza for germination and reproduction, and pest management, to name a few (Phillips et al., 2020).

To address these issues, an integrated strategy for orchid conservation and production has been reported to be essential (Kumar, 2012: Jalal, 2012). The main strategy remains to be in situ conservation of orchids. This entails identifying orchids and keeping them in their native environments, careful preservation of native forests with restoration initiatives, emphasis on threatened endemic species, establishment of orchid seed banks for genetic diversity, community awareness through education, and designated conservation areas for both tourists and academics. Another prominent method to ensure mass production of orchids is tissue culture technology (Lal et al. 2020). The technology offers the advantage of large-scale production of good quality, disease-free orchids under study in a limited space and time, ensuring continuous supply all year round. However, an efficient hardening and acclimatization protocol is essential for the survival of tissue-culture-raised orchids in field conditions. Besides clonal propagation, tissue culture studies on medicinal orchids provide the scope of the development of new varieties (through somaclonal variations or genetic transformation) and enhanced secondary metabolite production (through *in vitro* manipulations or hairy-root culture).

1.8 CONCLUSION

Orchids stand as emblematic symbols of biodiversity, cultural heritage, and medicinal treasure troves. From their ornamental allure to their therapeutic efficacy, these enigmatic plants have left an indelible mark on human history and continue to captivate researchers and enthusiasts alike. Orchids have played pivotal roles in traditional medicine and cultural practices across various civilizations. Chemical analyses have shed light on the diverse array of bio-active compounds found in orchids, from alkaloids to volatile organic compounds, contributing to their medicinal and aromatic properties. Recent studies on endangered orchid species in the Himalayan regions have underscored their potential as sources of active ingredients, further validating their historical usage in traditional medicine. Considering their paramount importance and burgeoning market value, concerted efforts are required towards their conservation and sustainable utilization. There is a need to integrate scientific research with traditional knowledge systems that can unlock the complete potential of orchids as sources of both aesthetic pleasure and therapeutic benefits. However, their conservation and sustainable utilization are paramount amidst rising demand and habitat degradation.

1.8.1 Acknowledgments

The authors extend their sincere gratitude to Mr. Mayuresh Prashant Kulkarni from Maharashtra, Ms. Veronika Làng from Hungary, Mr. Egidio Perantoni from Italy, Fahri Orchid from Indonesia, Mr. Kyle Pierce and Mr. Frank, who generously provided captivating photographs used in this chapter. We also highly value their interest and enthusiasm towards contributing to this academic venture.

REFERENCES

Acemi, A., Çobanoğlu, Ö., & Türker-Kaya, S. 2019. FTIR-based comparative analysis of glucomannan contents in some tuberous orchids, and effects of pre-processing on glucomannan measurement. *Journal of the Science of Food and Agriculture*, *99*(7): 3681–3686.

Adams, S.J., Kumar, T.S., & Muthuraman, G. 2023. Phytochemistry and biological activities of *Crepidium acuminatum* (D. Don) Szlach.: A systematic review. *Phytochemistry and Pharmacology of Medicinal Plants, 2.*

Babar, A., Naser, A., Saiba, S., Aftab, A., Shah, A.K., & Firoz, A. 2015. Essential oils used in aromatherapy: A systemic review. *Asian Pacific Journal of Tropical Biomedicine*, *5*(8): 601–611.

Behera, D., Rath, C.C., & Mohapatra, U. 2013. Medicinal orchids in India and their conservation: A review. *Floriculture and Ornamental Biotechnology*, *7*(1): 53–59.

Berliocchi, L. 2004. In Griffiths M, (ed.) *The orchid in lore and legend.* Timber Press.

Bhatnagar, J.K., Handa, S.S., & Duggal, S.C. 1971. Chemical investigations on *Microstylis wallichi. Planta Medica*, *20*(4): 156–161.

Bose, T.K., Bhattacharjee, S.K., Das, P., & Basak, U.C. 1999. *Orchids of India* (II). Naya Prokash.

Braunschmid, H., Guilhot, R., & Dötterl, S. 2020. Floral scent and pollinators of *Cypripedium calceolus* L. at different latitudes. Diversity 2021, 13: 5.

Dash, P.K., Sahoo, S., & Bal, S. 2008. Ethnobotanical studies on orchids of Niyamgiri hill ranges, Orissa, India. *Ethnobotanical Leaflets*, *2008*(1): 9.

D'Auria, M., Lorenz, R., Mecca, M., Racioppi, R., Antonio Romano, V., & Viggiani, L. 2020. Fragrance components of *Platanthera bifolia* subsp. osca and *Platanthera chlorantha* were collected in several sites in Italy. *Natural Product Research*, *34*(19): 2857–2861.

De, L.C. 2023. Pharmaceutical properties of orchids. *The Pharma Innovation Journal*, *12*(6): 1454–1460.

De, L.C., Khan, A.M., Kumar, R., & Medhi, R.P. 2014. Orchid farming—a remunerative approach for farmers' livelihood. *International Journal of Science and Research*, *3*: 468–471.

De, L.C., & Medhi, R.P. 2015. Orchid diversified components of farming systems for profitability and livelihood security of small and marginal farmers. *Journal of Global Biosciences*, *4*(2): 1393–1406.

De, L.C., & Pathak, P. 2015. Value addition in orchids. *Journal of Orchid Society of India*, *29*: 31–37.

De, L.C., Rao, A.N., Rajeevan, P.K., Promila, P., & Singh, D. 2015. Medicinal and aromatic orchids-an overview. *International Journal of Current Research*, *7*(9): 19931–19935. *https://doi.org/10.2478/9783110426403.12*

Fay, M.F. 2018. Orchid conservation: How can we meet the challenges in the twenty-first century? *Botanical Studies*, *59*: 1–6.

Fladby, T., Bryhn, G., Halvorsen, O., Rosé, I., Wahlund, M., Wiig, P., & Wetterberg, L. 2004. Olfactory response in the temporal cortex of the elderly measured with near-infrared spectroscopy: A preliminary feasibility study. *Journal of Cerebral Blood Flow & Metabolism*, *24*(6): 677–680.

Friesen, A., & Friesen, B. 2012. *The herbal power of orchids*. W. Zuckschwerdt Verlag.

Gam, D.T., Khoi, P.H., Ngoc, P.B., Linh, L.K., Hung, N.K., Anh, P.T.L., Thu, N.T., Hien, N.T.T., Khanh, T.D., & Ha, C.H. 2020. LED lights promote growth and flavonoid accumulation of *Anoectochilus roxburghii* and are linked to the enhanced expression of several related genes. *Plants*, *9*(10): 1344.

Gupta, R., Agarwal, M., & Baslas, R.K. 1978. Chromatographic separation and identification of various constituents of essential oil from the bulb of *M. accuminata*. Indian Perfume, 22(4): 287–288.

He, X., Wang, X., Fang, J., Zhao, Z., Huang, L., Guo, H., & Zheng, X. 2017. *Bletilla striata*: Medicinal uses, phytochemistry and pharmacological activities. *Journal of Ethnopharmacology*, *195*: 20–38.

Hossain, M.M. 2011. Therapeutic orchids: Traditional uses and recent advances—an overview. *Fitoterapia*, *82*(2): 102–140.

Huehne, P.S., Bhinija, K., Srisomsap, C., Chokchaichamnankit, D., Weeraphan, C., Svasti, J., & Mongkolsuk, S. 2020. Detection of superoxide dismutase (Cu–Zn) isoenzymes in leaves and pseudobulbs of *Bulbophyllum morphologlorum* Kraenzl orchid by comparative proteomic analysis. *Biochemistry and Biophysics Reports*, *22*: 100762.

Jalal, J.S. 2012. Status, threats and conservation strategies for orchids of western Himalaya, India. *Journal of Threatened Taxa*, *4*(15): 3401–3409.

Jalal, J.S., Kumar, P., & Pangtey, Y.P.S. 2008. Ethnomedicinal orchids of Uttarakhand, western Himalaya. Ethnobotanical Leaflets, 2008(1): 164.

Jersáková, J., Malinová, T., Jeřábková, K., & Dötterl, S. 2011. Biological flora of the British Isles: Pseudorchis albida (L.) Á. & D. Löve. *Journal of Ecology*, *99*(5): 1282–1298.

Kaiser, R. 1993. *The scent of orchids: Olfactory and chemical investigations*. Elsevier Science Publishers BV.

Kant, R., Verma, J., & Thakur, K. 2012. Distribution pattern, survival threats and conservation of 'Astavarga'Orchids in Himachal Pradesh, North-West Himalaya. *Plant Arch*, *12*(1): 165–168.

Kiziltas, H., Goren, A.C., Alwasel, S.H., & Gulcin, İ. 2022. Sahlep (*dactylorhiza osmanica*): Phytochemical analyses by LC-HRMS, molecular docking, antioxidant activity, and enzyme inhibition profiles. *Molecules*, *27*(20): 6907.

Kumar, P. (Ed.). 2012. Status, threats and conservation strategies for orchids of western Himalaya, India. *Journal of Threatened Taxa*, *4*(15): 3401–3409.

Lal, A., Pant, M., Kumar, A., Palni, L.M.S., Singh, A., Siddiqui, M.R., & Husen, A. 2023. Clonal propagation and hardening of epiphytic orchids (*Aerides multiflora* and *Rhyncostylis retusa*) for sustainable conservation and utilization. *South African Journal of Botany*, *162*: 402–411.

Lal, A., Pant, M., Palni, L.M.S., & Kumar, A. 2020. Development of rapid micropropagation protocol for germplasm conservation of two orchid species–*Aerides multiflora* Roxb. and *Rhynchostylis retusa* (L.) Blume. Asian Journal of Conservation Biology, 9(2): 341–347.

Lubinsky, P., Bory, S., Hernández Hernández, J., Kim, S.C., & Gómez-Pompa, A. 2008. Origins and dispersal of cultivated vanilla (*Vanilla planifolia* Jacks.[Orchidaceae]). *Economic botany*, *62*: 127–138.

Lynn, M., Flam, M., Doughlas-Paxton, D., & Raymond, J. 1992. Improved food intake and reduced nausea and vomiting in patients given a restricted diet while receiving cisplatin chemotherapy. *Journal of the American Dietetic Association*, *92*(1): 58–61.

Medhi, R.P., & Chakrabarti, S. 2009. Traditional knowledge of NE people on conservation of wild orchids.
Mokni, R.E., Hammami, S., Dall'Acqua, S., Peron, G., Faidi, K., Braude, J.P., Sebei, H., & Aouni, M.H.E. 2016. Chemical composition, antioxidant and cytotoxic activities of essential oil of the inflorescence of *Anacamptis coriophora* subsp. fragrans (Orchidaceae) from Tunisia. *Natural Product Communications*, *11*(6): 1934578X1601100640.
Mudoi, K.D., Borah, P., Gorh, D., Gupta, T., Sarmah, P., Bhattacharjee, S., Roy, P., & Saikia, S.P. 2023. Biotechnological interventions and societal impacts of some medicinal orchids. In *Advances in orchid biology, biotechnology and omics*, 59–144. Springer Nature.
Naczk, A.M., Kowalkowska, A.K., Wiśniewska, N., Haliński, Ł.P., Kapusta, M., & Czerwicka, M. 2018. Floral anatomy, ultrastructure and chemical analysis in *Dactylorhiza incarnata/maculata* complex (Orchidaceae). *Botanical Journal of the Linnean Society*, *187*(3): 512–536.
Natta, S., Mondol, M.S.A., Pal, K., Mandal, S., Sahana, N., Pal, R., Pandit, G.K., Alam, B.K., Das, S.S., Biswas, S.S., & Kalaivanan, N.S. 2022. Chemical composition, antioxidant activity and bioactive constituents of six native endangered medicinal orchid species from north-eastern Himalayan region of India. *South African Journal of Botany*, *150*: 248–259.
Pal, R., Meena, N.K., Dayamma, M., & Singh, D.R. 2022. Ethnobotany and recent advances in Indian medicinal orchids. *Orchids Phytochemistry, Biology and Horticulture: Fundamentals and Applications*: 361–387.
Panda, A.K., & Mandal, D. 2013. The folklore medicinal orchids of Sikkim. *Ancient Science of Life*, *33*(2): 92.
Pant, B. 2013. Medicinal orchids and their uses: Tissue culture a potential alternative for conservation. *African Journal of Plant Science*, *7*(10): 448–467.
Pellegrino, G., Bellusci, F., & Musacchio, A. 2008. Double floral mimicry and the magnet species effect in dimorphic co-flowering species, the deceptive orchid *Dactylorhiza sambucina* and rewarding *Viola aethnensis. Preslia*, *80*: 411–422.
Peng, J., Zhan, L., Dong, F., Han, X., Qi, Y., Xu, Y., Xu, L., Xu, Q., & Liu, K. 2008. A comparative study of chromatographic methods for separating chemical compounds from *Spiranthes australis* (R. Brown) Lindl roots. *Separation and purification technology*, *59*(3): 262–269.
Phillips, R.D., Reiter, N., & Peakall, R. 2020. Orchid conservation: From theory to practice. *Annals of Botany*, *126*(3): 345–362.
Pridgeon, A.M., Cribb, P.J., Chase, M.W., & Rasmussen, F. 2001. Genera Orchidacearum, Vol. II: Orchidoideae (Part I).
Robustelli della Cuna, F.S., Cortis, P., Esposito, F., De Agostini, A., Sottani, C., & Sanna, C. 2022. Chemical composition of essential oil from four sympatric orchids in NW-Italy. *Plants*, *11*(6): 826.
Salzmann, C.C., Nardella, A.M., Cozzolino, S., & Schiestl, F.P. 2007. Variability in floral scent in rewarding and deceptive orchids: The signature of pollinator-imposed selection?. *Annals of Botany*, *100*(4): 757–765.
Sandrasagaran, U.M., Subramaniam, S., & Murugaiyah, V. 2014. New perspective of *Dendrobium crumenatum* orchid for antimicrobial activity against selected pathogenic bacteria. *Pakistan Journal of Botany*, *46*(2): 719–724.
Sezik, E. 2002. Turkish orchids and salep. *Acta Pharmaceutica Turcica*, *44*(3): 151–157.
Shapoo, G.A., Kaloo, Z.A., Ganie, A.H., & Singh, S. 2013. Ethnobotanical survey and documentation of some orchid species of Kashmir Himalaya, J&K-India. *International Journal of Pharmaceutical and Biological Research*, *4*(2): 32–40.
Sharma, A., & Pathak, P. 2020. The budding potential of orchids in the cosmeceutical sector: Role of orchids in skincare and health. *Journal of Orchid Society of India*, *34*: 79–85.
Štajner, D., Popović, B.M., Kapor, A., Boža, P., & Štajner, M. 2010. Antioxidant and scavenging capacity of *Anacamptis pyrimidalis* L.–pyrimidal orchid from Vojvodina. *Phytotherapy Research: An International Journal Devoted to Pharmacological and Toxicological Evaluation of Natural Product Derivatives*, *24*(5): 759–763.
Su, Z., Yang, Y., Chen, S., Tang, Z., & Xu, H. 2023. The processing methods, phytochemistry and pharmacology of *Gastrodia elata* Bl.: A comprehensive review. *Journal of Ethnopharmacology*: 116467.
Ticktin, T., Charitonidou, M., Douglas, J., Halley, J.M., Hernández-Apolinar, M., Liu, H., Mondragón, D., Pérez-García, E.A., Tremblay, R.L., & Phelps, J. 2023. Wild orchids: A framework for identifying and improving sustainable harvest. Biological Conservation, 277: 109816.
Vaidya, B.N. 2019. Nepal: A global hotspot for medicinal orchids. *Medicinal Plants: From Farm to Pharmacy*: 35–80.

Wróblewska, A., Szczepaniak, L., Bajguz, A., Jędrzejczyk, I., Tałałaj, I., Ostrowiecka, B., Brzosko, E., Jermakowicz, E., & Mirski, P. 2019. Deceptive strategy in *Dactylorhiza* orchids: Multidirectional evolution of floral chemistry. *Annals of botany*, *123*(6): 1005–1016.

Wu, Y.B., Peng, M.C., Zhang, C., Wu, J.G., Ye, B.Z., Yi, J., Wu, J.Z., & Zheng, C.J. 2020. Quantitative determination of multi-class bioactive constituents for quality assessment of ten *Anoectochilus*, four *Goodyera* and one *Ludisia* species in China. *Chinese Herbal Medicines*, *12*(4): 430–439.

Zhan, H.D., Zhou, H.Y., Sui, Y.P., Du, X.L., Wang, W.H., Dai, L., Sui, F., Huo, H.R., & Jiang, T.L. 2016. The rhizome of *Gastrodia elata* Blume–An ethnopharmacological review. *Journal of ethnopharmacology*, *189*: 361–385.

2 Conservation of Medicinal Orchids through Plant Tissue Culture

Prasenjit Paul and Suman Kumaria

2.1 INTRODUCTION

Medicinal plants are a significant part of the plant resources and have a major influence on human health. Orchids are also known for their medicinal values, especially in traditional systems of medicine. The primary reason for these plants' therapeutic significance is their ability to produce a variety of chemical compounds that have physiological effects on treating a wide variety of illnesses. Orchids have been used medicinally since the Vedic era. Four species of orchids—*Malaxis muscifera*, *M. acuminata*, *Habenaria intermedia*, and *H. edgeworthi*—combine to form "Ashtawarga," a set of eight medications used in Ayurvedic medicine to prepare tonics like "Chyavanparas" (Jalal et al., 2008). From as early as 200 BC, the Chinese pharmacopeia "Sang Nueng Pen Tsao Ching" emphasized the importance of dendrobes as a source of tonic, analgesic, astringent, and anti-inflammatory chemicals (Singh and Tiwari, 2007). Additionally, one of Traditional Chinese Medicine's (TCM) most valuable classes is *Herba Dendrobii*, which consists of the stems of several dendrobes. *Herba Dendrobii* is divided into two major categories, known as "*Fengdou Shihu*"and "*Huangcao Shihu,*" based on the morphological traits and processing techniques of the orchids used in the preparations (Yuan et al., 2011). The most common kind of *Herba Dendrobii* is called *Hunagcao Shihu*, and it is widely used for treating respiratory and lung conditions like coughs and colds and boosting immunity (Kuang et al., 2005; Zhang et al., 2005). Some of the bioactive substances, which are claimed to be present in a wide variety of plants, including orchids, comprise alkaloids, tannins, flavonoids, and phenolic compounds (Bhattacharyya et al., 2015; Paul and Kumaria, 2020). Along with these bioactive components, other substances with medicinal values include ß-sitosterol, jibantine, aeredin, agrostophyllinol, arundinan, moscatilin, dendrobine, and rotundatin reported to be present in different species of orchids (Singh and Duggal, 2009; Paul et al., 2017). Some of the medicinally important orchids are *Cymbidium lowianum, Dendrobium fimbriatum*, *Coelogyne fimbriata*, *Dendrobium densiflorum*, *Paphiopedilum insigne*, *Dendrobium chrysotoxum*, *Dendrobium jenkinsii*, *Rhynchostylis retusa,* and *Dendrobium chrysanthum* (Figure 2.1).

Currently, the International Union for Conservation of Nature and Natural Resources' (IUCN) Red Data Book lists all species of orchids (Singh, 2001). Various anthropogenic factors, such as habitat destruction, degradation, fragmentation, and illicit collection for sale and consumption, are accountable for the decline to low levels in wild populations of medicinal orchids (Pant, 2013). As a result, the swift decimation of medicinally significant orchids from their native habitats necessitates prompt action to ensure the survival of these significant members of the plant kingdom. *In vitro* technologies allow the contemporary instruments of biotechnology to directly support plant conservation initiatives (Tandon and Kumaria, 1998, 2005).

DOI: 10.1201/b23375-2

FIGURE 2.1 Some medicinal orchids: a) *Cymbidium lowianum,* b) *Dendrobium fimbriatum,* c) *Coelogyne fimbriata,* d) *Dendrobium densiflorum,* e) *Paphiopedilum insigne,* f) *Dendrobium chrysotoxum,* g) *Dendrobium jenkinsii,* h) *Rhynchostylis retusa,* i) *Dendrobium chrysanthum.*

2.2 TISSUE CULTURE OF MEDICINAL ORCHIDS

Plant tissue culture (PTC) methods have been a practical choice for the commercial cultivation of plants for *ex-situ* conservation, reforestation, and the extraction of medicinal compounds (Espinosa-Leal et al., 2018). Comparing PTC to traditional field culture, PTC has a clear advantage as it does not rely on seasonal, geographical, or environmental fluctuations, allowing for year-round, uninterrupted production (Debnath et al., 2006; Paul et al., 2017). Techniques towards *in vitro* production are currently indispensable for the synthesis of bioactive substances and secondary metabolites with economic potential. Numerous researchers have thoroughly demonstrated the ability of *in vitro* procedures to promote the aggregation of numerous secondary metabolites with therapeutic value as well as the antioxidant potential of the cultured tissues (Verpoorte et al., 2002; Paul et al., 2017; Bhattacharyya et al., 2022). Incorporating precursors and intermediates in the culture to improve secondary metabolites has become a realistic alternative as demand for these compounds rises. Adding biochemical precursors exogenously to cultures of orchids has been shown to increase the production of secondary metabolites in these plants (Paul and Kumaria, 2020).

Ex-situ conservation of medicinal orchids can be achieved using different plant tissue culture strategies *viz*., asymbiotic seed culture for embryo rescue, meristem culture, anther culture, ovary culture, protoplast culture, etc.

2.2.1 Asymbiotic Seed Culture

The embryo inside orchid seeds is underdeveloped as there is no endosperm, which limits the embryo's access to direct nutrients. As a result, the long-term conservation of numerous medicinally significant orchids has benefitted from asymbiotic seed germination utilizing tissue culture techniques. In a number of medicinal orchids, including *Ansellia africana, Dendrobium fimbriatum, D. hookerianum, D. moniliforme, D. longicornu, D. formosum, Cymbidium eburneum, Paphipedilum insigne, Vanda cristata,* and *Vanda pumila*, asymbiotic seed germination has been successfully accomplished on different nutrient media *viz*., MS, Knudson C, Gamborg B_5, etc. (Kumaria and Tandon, 1991; Dohling et al., 2008; Gogoi et al., 2012; Paul et al., 2012; Bhattacharyya et al., 2017; Diengdoh et al., 2017; Utami and Hariyanto, 2020; Liu et al., 2023; Pathak et al., 2023). Nutrient media are essential for the germination of orchid seeds because they are rich in micronutrients, macronutrients, vitamins, carbon sources, plant growth regulators (PGRs), etc. The physiological needs of macro- and micronutrients are crucial to the growth and development of plants. An integral part of a balanced nutrient medium is the proper source of nitrogen. Stenberg and Kane (1998) noted the beneficial effects of nitrogen, which the orchid seeds may effectively use during early germination and morphogenesis. Plants also need vitamins as catalysts for a variety of metabolic processes necessary for their optimal growth. In PTC, plant growth regulators like auxin, cytokinin, gibberellins, and abscisic acid are essential because they regulate processes like stem elongation, tropism, apical dominance, and others (Saad and Elshahed, 2012). The method of *in vitro* seed germination has also assisted in rescuing a significant proportion of orchid embryos.

2.2.2 Meristem Culture

Meristem culture, which has been widely utilized to propagate a variety of medicinal plants, including orchids, is another component of PTC. The process of meristem culture involves removing the meristematic portion of plants, such as the tips of the shoots or roots, and placing it in an appropriate growth medium that contains minerals, vitamins, and plant hormones. In cultures, it promotes tissue development and cell division. Usually, propagating plants from meristem explants offers a means of producing plants that are devoid of viruses and other systemic diseases. A further benefit of meristem culture is its intrinsic ability to maintain genetic stability in micropropagated plants. Meristem culture's advantages have opened up exciting opportunities for orchid propagation (Arditti and Ernst, 1993). The meristem culture approach has been used to successfully micropropagate a variety of medicinally significant orchids, including *Dendrobium densiflorum, D. fimbriatum,* and *Vanda coerulea* (Seeni and Latha, 2000; Roy and Banerjee, 2003; Pradhan et al., 2013). An economically noteworthy orchid genera with strong therapeutic properties, *Cymbidium*, often known as "boat orchids," has been successfully conserved utilizing PTC technology (Balilashaki et al., 2023).

2.2.3 Anther Culture or Androgenesis

In addition to other PTC applications, anther culture has contributed to the conservation of orchids. Androgenesis, also known as anther culture, is a peculiar biological phenomenon that has been found to occur in angiosperms. It is the process of stopping the pollen grains' growth and directing them onto a somatic path, and it has wide applications in the production of haploids. The valuable medicinal orchid, *Rhynchostylis retusa*, also known as the "fox tail orchid" and a member of the Orchidaceae family has been propagated via anther culture (Sharma, 2012). Using anther culture, *Phalaenopsis*, a significant member of the orchidaceae family, has also been successfully micropropagated (Kornienko and Tesliuk, 2021).

2.2.4 Ovary Culture or Gynogenesis

Ovary culture or gynogenesis is the process of culturing ovaries in a suitable growth medium under defined growth conditions that have been removed from either pollinated or unpollinated flowers. Ovary culture has aided in the research of embryo development, overcoming incompatibility, and preventing precocious floral abscission. An important medicinal orchid, *Dendrobium nobile,* utilized in traditional herbal medicine, has been successfully cultured using ovary culture (Ito, 1967). Also, *Gymnadenia conopsea*, a terrestrial orchid, has been successfully grown through ovary culture and is traditionally utilized as a major component of Tibetan and Mongolian medicine (Pindel and Pindel, 2004). The treasured European folk medicinal plant, the terrestrial orchid *Epipactis helleborine*, has also been effectively propagated via ovary culture (Pindel and Pindel, 2004; Pant, 2013). *Cymbidium*, a genus of highly prized medical orchids, has effectively produced seeds and protocorms *in vitro* using ovary culture (Kato et al., 2018).

2.2.5 Suspension Culture

The conservation of medicinally significant orchids through suspension culture is another crucial component of PTC, which not only leads to enhanced production of propagules but also is beneficial for increasing the accumulation of secondary metabolites. The chemical underpinnings of medicinal orchids' therapeutic properties come from their secondary metabolites. To boost the formation of secondary metabolites in PTC, precursors and intermediates can be introduced to the growth medium (Paul and Kumaria, 2020). In addition to enhancing the production of secondary metabolites, PTC can serve as a consistent source for the synthesis of secondary metabolites. One of the treasured medicinal orchids, *Malaxis acuminata*, has been effectively preserved via suspension culture and has found widespread applications in numerous herbal and nutraceutical preparations (Bhattacharyya et al., 2023). An epiphytic sympodial orchid, *Dendrobium fimbriatum,* which is one of the many therapeutic orchids used worldwide in herbal medicines, has also been propagated through suspension culture (Paul and Kumaria, 2020). Another rare medicinal orchid from the Himalayas called *Habenaria edgeworthii* has been grown successfully utilizing callus suspension cultures (Giri et al., 2012).

2.2.6 Protoplast Culture

The culture of protoplasts, which has numerous applications in the field of medicinal plant conservation, is another significant method of PTC (Arditti, 2009). Naked plant cells, or plant cells devoid of a cell wall, are referred to as protoplasts. Any plant part, including the root, stem, leaves, or embryo, can be utilized to isolate protoplasts, which are then cultivated in artificial media under an artificial environment to facilitate cell division and plant regeneration. The intriguing aspect of protoplast culture in plant tissue culture offers opportunities for the creation of hybrids and cybrids, transplantation of nuclei, and the development of genetic transformants. One of the most significant orchidaceae members, noted for its medicinal properties, is the *Dendrobium* genus. With the use of the protoplast culture method, *Dendrobium Sonia* has been successfully multiplied (Khentry et al., 2006). The moth orchid, or *Phalaenopsis* species, is one of the most promising sources of therapeutic compounds that have been productively propagated by protoplast culture (Ichihashi and Shigemura, 2005; Shrestha et al., 2007). *Oncidium*, often known as "Golden Showers" or "Dancing Ladies," is another medicinally noteworthy orchid that has also been established using this method (Choo, 1992).

2.2.7 Cryopreservation

Cryopreservation, which typically involves the preservation of biological materials in liquid nitrogen at -196° C or its vapor phase at -150°C, is a further crucial method used in PTC. The germplasm of plant species that are vegetatively propagated or that have non-viable or difficult-to-germinate seeds may only currently be preserved through this method. *Paphiopedilum insigne*, a notable medicinal orchid known as the "Lady Slipper Orchid," has been effectively cryopreserved utilizing the vitrification and encapsulation-vitrification approaches (Friesen and Friesen, 2012; Diengdoh et al., 2019). Compared to conventional freezing procedures, vitrification has various advantages. It decreases the time needed for the cryopreservation process, high-concentration cryoprotectants shorten the cryoprotectant's exposure period, limit penetration damage, and do away with the expense of slow-freezing equipment and its upkeep. Encapsulation-vitrification, on the other hand, combines the benefits of encapsulation-dehydration (easiness of handling of encapsulated explants) with vitrification (rapidity of application). Numerous medicinal orchids, including *D. nobile, Bletilla striata, B. formosana, Calanthe masuca*, etc., have reported success with cryopreservation (Ishikawa et al., 1997; Mohanty et al., 2012; Hu et al., 2013; Decruse et al., 2023).

2.3 CONCLUSION

The demand for medicinal orchids is very high in the international market. Natural populations of medicinal orchids are being stressed and are dwindling as a result of rising demand spurred on by the growing population. To compensate for such declines, plant tissue culture has made significant contributions to the conservation of a sizable portion of the therapeutic orchids'. Plant tissue culture's standout benefit is that it aids in the quicker and more reliable conservation of plant material by retaining the same clone or stock of plant material to a greater extent. Tissue culture, in addition to being a trustworthy conservation technique, also helps to be a source of active and enhanced secondary metabolite production in a controlled growth environment.

REFERENCES

Arditti, J. 2009. *Micropropagation of orchids*. Oxford, UK: Blackwell Publishing.

Arditti, J., & Ernst, R., 1993. *Micropropagation of orchids*. New York: John Wiley & Sons, Inc.

Balilashaki, K., Martinez-Montero, M. E., Vahedi, M., Cardoso, J. C., Silva Agurto, C. L., Leiva-Mora, M., Feizi, F., & Musharof Hossain, M. 2023. Medicinal use, flower trade, preservation and mass propagation techniques of *Cymbidium* orchids—An Overview. *Horticulturae*, 9(6): 690.

Bhattacharyya, P., Kumar, S., Lalthafamkimi, L., Sharma, R., Kumar, D., Singh, D., & Kumar, S. 2023. Molecular and phytomedicinal stability of long term micropropagated *Malaxis acuminata*: An endangered terrestrial orchid of biopharmaceutical importance. *South African Journal of Botany*, 155: 372–382.

Bhattacharyya, P., Kumar, V., & Van Staden, J. 2017. Assessment of genetic stability amongst micropropagated *Ansellia africana*, a vulnerable medicinal orchid species of Africa using SCoT markers. *South African Journal of Botany*, 108: 294–302.

Bhattacharyya, P., Kumaria, S., Job, N., & Tandon, P. 2015. Phyto-molecular profiling and assessment of antioxidant activity within micropropagated plants of *Dendrobium thyrsiflorum*: A threatened, medicinal orchid. *Plant Cell, Tissue and Organ Culture*, 122: 535–550.

Bhattacharyya, P., Lalthafamkimi, L., Sharma, T., Kumar, D., Kumar, A., Kumar, D., & Kumar, S. 2022. Metabolic and tissue-specific expression profiling in micropropagated plants of *Malaxis acuminata*: An endangered medicinal orchid. *Plant Cell, Tissue and Organ Culture*, 151(3): 535–549.

Choo, A. K. M. 1992. *Protoplast and cell culture of certain orchids*. https://scholarbank.nus.edu.sg/handle/10635/171932

Debnath, M., Malik, C. P., & Bisen, P. S. 2006. Micropropagation: A tool for the production of high quality plant-based medicines. *Current Pharmaceutical Biotechnology*, 7(1): 33–49.

Decruse, S. W., Ajeeshkumar, S., & Vargheese, S. 2023. Pollinia cryopreservation of selected orchid species of Western Ghats, India. In *Pollen Cryopreservation Protocols*. New York: Springer US.

Diengdoh, R. V., Kumaria, S., & Das, M. C. 2019. Antioxidants and improved regrowth procedure facilitated cryoconservation of *Paphiopedilum insigne* Wall. Ex. Lindl.-An endangered slipper orchid. *Cryobiology*, 87: 60–67.

Diengdoh, R. V., Kumaria, S., Tandon, P., & Das, M. C. 2017. Asymbiotic germination and seed storage of *Paphiopedilum insigne*, an endangered lady's slipper orchid. *South African Journal of Botany*, 112: 215–224.

Dohling, S., Kumaria, S., & Tandon, P. 2008. Optimization of nutrient requirements for asymbiotic seed germination of *Dendrobium longicornu* Lindl. and *D. formosum* Roxb. *Proceedings Indian National Science Academy*, 74(4): 167–171.

Espinosa-Leal, C. A., Puente-Garza, C. A., & García-Lara, S. 2018. *In vitro* plant tissue culture: Means for production of biological active compounds. *Planta*, 248: 1–18.

Friesen, A., & Friesen, B. 2012. *Orchids in ethnobotany and ethnomedicine*. The herbal power of orchids, 4: 18–24.

Giri, L., Dhyani, P., Rawat, S., Bhatt, I. D., Nandi, S. K., Rawal, R. S., & Pande, V. 2012. *In vitro* production of phenolic compounds and antioxidant activity in callus suspension cultures of *Habenaria edgeworthii*: A rare Himalayan medicinal orchid. *Industrial Crops and Products*, 39: 1–6.

Gogoi, K., Kumaria, S., & Tandon, P. 2012. *Ex situ* conservation of *Cymbidium eburneum* Lindl.: A threatened and vulnerable orchid, by asymbiotic seed germination. *3 Biotech*, 2: 337–343.

Hu, W. H., Yang, Y. H., Liaw, S. I., & Chang, C. 2013. Cryopreservation the seeds of a Taiwanese terrestrial orchid, *Bletilla formosana* (Hayata) Schltr. by vitrification. *Botanical Studies*, 54: 1–7.

Ichihashi, S., & Shigemura, S. 2005. *Phalaenopsis* and protoplast culture. In *Proceedings of the 17th World Orchid Conference: "Sustaining Orchids for the Future" 2002*, Selangor (Malaysia), Natural History Publications(Borneo) Sdn Bhd, 257–261.

Ishikawa, K., Harata, K., Mii, M., Sakai, A., Yoshimatsu, K., & Shimomura, K. 1997. Cryopreservation of zygotic embryos of a Japanese terrestrial orchid (*Bletilla striata*) by vitrification. *Plant Cell Reports*, 16: 754–757.

Ito, I. 1967. *In vitro* culture of ovary in orchids (II): Influence of vitamins upon the growth of ovary of *Dendrobium nobile* and other several problems (Agriculture). The scientific reports of Kyoto Prefectural University. *Agriculture*, 19: 6–12.

Jalal, J. S., Kumar, P., & Pangtey, Y. P. S. 2008. Ethnomedicinal orchids of Uttarakhand, western Himalaya. *Ethnobotanical Leaflets*, 2008(1): 164.

Kato, J., Hirai, R., Iguchi, K., Iwata, A., & Ichihashi, S. 2018. Seed and protocorm development by *in vitro* ovary culture at the fertilization phase in *Cymbidium*. *Acta Horticulturae,* 1262: 141–148.

Khentry, Y., Paradornuvat, A., Tantiwiwat, S., Phansiri, S., & Thaveechai, N. 2006. Protoplast isolation and culture of *Dendrobium* Sonia "Bom 17". *Agriculture and Natural Resources*, 40(2): 361–369.

Kornienko, A. A., & Tesliuk, N. I. 2021. Selection efficiency of different initial explants types of *Phalaenopsis* orchids species. *Editorial Board*, 37.

Kuang, L., Zhang, K., & Commission, C. P., 2005. *Pharmacopoeia of the People's Republic of China*. Beijing: People's Medical Publishing House.

Kumaria, S., & Tandon, P. 1991. Asymbiotic germination of *Dendrobium fimbriatum* var. *oculatum* Hk. f. seeds on different media. *Proceedings of the Indian National Science Academy. Part B: Biological Sciences*, 57(3–4): 277–279.

Liu, X., Sun, L., Nie, T., Chen, Y., & Yin, Z. 2023. *In vitro* rapid propagation technology system of *Dendrobium moniliforme* (L.) Sw., a threatened orchid species in China. *Plant Biotechnology Reports*, 1–10.

Mohanty, P., Das, M. C., Kumaria, S., & Tandon, P. 2012. High-efficiency cryopreservation of the medicinal orchid *Dendrobium nobile* Lindl. *Plant Cell, Tissue and Organ Culture*, 109: 297–305.

Pant, B. 2013. Medicinal orchids and their uses: Tissue culture a potential alternative for conservation. *African Journal of Plant Science*, 7(10): 448–467.

Pathak, P., Kumari, A., Chandler, B. D., & Zettler, L. W. 2023. *In vitro* propagation and phytochemical analysis of *Vanda cristata* Wall. ex Lindl: An endangered medicinal orchid of biopharmaceutical importance. *South African Journal of Botany*, 153: 109–123.

Paul, P., Joshi, M., Gurjar, D., Shailajan, S., & Kumaria, S. 2017. *In vitro* organogenesis and estimation of β-sitosterol in *Dendrobium fimbriatum* Hook.: An orchid of biopharmaceutical importance. *South African Journal of Botany*, 113: 248–252.

Paul, P., & Kumaria, S. 2020. Precursor-induced bioaccumulation of secondary metabolites and antioxidant activity in suspension cultures of *Dendrobium fimbriatum*, an orchid of therapeutic importance. *South African Journal of Botany*, 135: 137–143.

Paul, S., Kumaria, S., & Tandon, P. 2012. An effective nutrient medium for asymbiotic seed germination and large-scale *in vitro* regeneration of *Dendrobium hookerianum*, a threatened orchid of northeast India. *AoB Plants*: plr032(doi:10.1093/aobpla/plr032).
Pindel, A., & Pindel, Z. 2004. Initiation of *in vitro* cultures of chosen endangered European species of orchids. *Folia Horticulturae*, 16(2): 111–117.
Pradhan, S., Paudel, Y. P., & Pant, B. 2013. Efficient regeneration of plants from shoot tip explants of *Dendrobium densiflorum* Lindl., a medicinal orchid. *African Journal of Biotechnology*, 12(12): 1378–1383.
Roy, J., & Banerjee, N. 2003. Induction of callus and plant regeneration from shoot-tip explants of *Dendrobium fimbriatum* Lindl. var. *oculatum* Hk. f. *Scientia Horticulturae*, 97(3–4): 333–340.
Saad, A. I., & Elshahed, A. M. 2012. Plant tissue culture media. *Recent Advances in Plant in Vitro Culture*, 30–40.
Seeni, S., & Latha, P. G. 2000. *In vitro* multiplication and ecorehabilitation of the endangered Blue Vanda. *Plant Cell, Tissue and Organ Culture*, 61: 1–8.
Sharma, V. 2012. Orchid micropropagation: Regeneration competence of anther culture. *Journal of Biotechnology and Biomaterials*, 2: 149.
Shrestha, B. R., Tokuhara, K., & Mii, M. 2007. Plant regeneration from cell suspension-derived protoplasts of *Phalaenopsis*. *Plant Cell Reports*, 26: 719–725.
Singh, A., & Duggal, S. 2009. Medicinal orchids-an overview. *Ethnobotanical Leaflets*, 2009(3): 3.
Singh, A. K. R., & Tiwari, C. 2007. Harnessing the economic potential of Orchids in Uttaranchal. *ENVIS Bulletin Himalayan Ecology*, 14: 1–3.
Singh, D. K. 2001. Orchid diversity in India: An overview. In Pathak, P., Sehgal, R. N., Shekhar, N., Sharma, M., & Sood, A. (eds) *Orchids: Science and Commerce*. 35–65.
Stenberg, M. L., & Kane, M. E. 1998. *In vitro* seed germination and greenhouse cultivation of *Encyclia boothiana* var. *Erythronioides*, an endangered Florida orchid. *Lindleyana-West Palm Beach*, 13: 101–112.
Tandon, P., & Kumaria, S. 1998. Threats to plant diversity in high altitude of North-East India and conservation of rare and endangered plants using biotechnological approaches. In *Science at High Altitude. Proceedings of the National Symposium*. India: Allied Publishers Ltd, 140–147.
Tandon, P., & Kumaria, S. 2005. Prospects of plant conservation biotechnology in India with special reference to Northeastern region. In Tandon, P., Sharma, M. and Swarup, R. (eds) *Biodiversity: Status and Prospects*. New Delhi: Narosa Publshing House, 79–91.
Utami, E. S. W., & Hariyanto, S. 2020. Organic compounds: Contents and their role in improving seed germination and protocorm development in orchids. *International Journal of Agronomy*, 1–12.
Verpoorte, R., Contin, A., & Memelink, J. 2002. Biotechnology for the production of plant secondary metabolites. *Phytochemistry Reviews*, 1: 13–25.
Yuan, Y. H., Hou, B. W., Xu, H. J., Luo, J., & Ding, X. Y. 2011. Identification of the geographic origin of *Dendrobium thyrsiflorum* on Chinese herbal medicine market using trinucleotide microsatellite markers. *Biological and Pharmaceutical Bulletin*, 34(12): 1794–1800.
Zhang, Y. B., But, P. P. H., Wang, Z. T., & Shaw, P. C. 2005. Current approaches for the authentication of medicinal *Dendrobium* species and its products. *Plant Genetic Resources*, 3(2): 144–148.

3 Conservation of Endangered Medicinal Orchid Species through Post-Harvest Handling for Future Sustainable Use

Manuj Awasthi, Abhishek Chandra, and Sunil Malik

3.1 INTRODUCTION

Orchids are considered to be the most abundant and alluring plant species to be found in various terrains of the world, from high mountains to the tropics (White and Sharma, 2000). Most orchids are found in the rainforests of Central and South America, India, Sri Lanka, Burma, Southern China, Thailand, Malaysia, Philippines, New Guinea, and Australia. *Cattleya, Laelia, Cymbidium* sp., *Vanda* sp., *Dendrobium,* and other orchids have played an important role in the development of today's beautiful hybrid orchids, which number more than 2,00,000. Orchids are the largest flower family, with 600 to 800 genera and 25,000 to 35,000 species. They are appreciated for their large flower variety, beautiful colors, and attractiveness as well as the fact that they keep well for up to 10 weeks. In the international floriculture market, among the top ten cut flowers, orchids rank in the sixth position, and among orchids, *Cymbidium* orchids are ranked at the top, and among floricultural crops, account for 3% of the total cut flower production. In addition to their extraordinary aesthetic and wider blooming span, orchids are a prominent economic indication, notably in horticulture and floristry, but they are also pertinent in the perfume and pharmaceutical industries (Pant and Swar, 2013). The conservation status showed that 61 % of species are vulnerable, 31% are endangered, and 6% are near to threatened. Orchids are the most fascinating and variegated group of monocots in under angiosperms. They are mainly cultivated for their extraordinary flowers, mainly used in floral adornments and arrangements and gardening, but very few for their medicinal properties. "*Jivanti* and *Rasna*" are the orchids used in Ayurvedic drug concoctions for several centuries. In India, there are 1,331 species of orchids belonging to 158 genera, and they grow up to an altitude of 5,000 meters. Indian terrestrial life is found in much humus with moist soil in the shade of trees in northwestern India. There are small flower orchids in the Western Ghats. Epiphytic orchids are found in Northeastern India, growing from sea levels to high altitudes.

Nowadays, approximately 50 species of orchids are broadly utilized for medication. A few of the species, like *Vanda roxburghii, Orchis latifolia*, and *D. nobile* have been recorded for their demonstrated therapeutic values. The most common phytochemicals reported in orchid species are alkaloids, triterpenoids, flavonoids, and stilbenoids. *Vanilla planifolia* is one of the important orchids as a source plant for Vanillin, which is used as a food flavoring agent. Orchids have been utilized in numerous herbal formulations since time immemorial (Table 3.1). The efficient utilization of this exceptional resource as therapeutics is not commensurate with their potential, mainly owing to their endangered type of orchids. Harnessing their wild populations for commercial purposes is not a sustainable solution. In vitro cell and tissue cultures, therefore, prove to be an alternative and

 DOI: 10.1201/b23375-3

FIGURE 3.1 Flower collection of different orchid species for in vitro pollen cryopreservation. (a) (b) and (c) Flower collection of Cymbidium species of orchid for extracting the pollen (d) Flower collection of Dendrobium species of orchid for extracting the pollen.

beneficial approach. The compositions of the culture media and the culture conditions have a significant impact on the physiological status of cells or tissues (Kaur et al., 2023).

Orchids, apart from their ornamental value, also possess medicinal properties which have been used in traditional medicine systems. The high demand for orchids in pharmaceutical and ornamental industries has led to the over-exploitation of orchids (Figure 3.1), which has led to the extinction or depletion of some species while some species are declared endangered (Mudoi et al., 2023).

Deforestation has become a major cause of the extinction of orchids. Conventional orchid propagation methods cannot overcome the challenges of slow growth and low seed germination rate, and they are prone to pests, diseases, and lack of nutrition. To deal with these challenges,

TABLE 3.1
Chemical Constituents and Uses of Some Rare Medicinal Orchids

Sl. No.	Name	Chemical Constituents	Ethnomedical Uses
1.	*Coelogyne cristata*	Coeloginanthrin, Coeloginanthridin, Combretastatin C-1, Coelogin	Used as an aphrodisiac and to cure wounds, boils, sores, and constipation
2.	*Coelogyne flaccida*	Flaccidin, Flaccidinin, Oxoflaccidin, Isooxoflaccidin	Treatment of headache, fever, and indigestion
3.	*Coelogyne flavida*	Flavidin, Flavidinin	Traditional medicines
4.	*Coelogyne ochracea*	Ochrone-A, Bochrolic acid, Ochrolon	Used in fractured bones
5.	*Coelogyne ovalis*	2,7- dihydroxy 3,4,6—tri methoxy 9,10- dihydrophenanthrne	Used as an aphrodisiac and to cure cough, urinary infections, and eye disorders
6.	*Cymbidium aloifolium*	Aloifol-I, Cymbinodin-A, B	Used as purgative and tonic to cure rheumatisms, nervous disorders, wounds, sores, boils, and fevers
7.	*Cymbidium giganteum*	Gigantol	Used in blood clotting
8.	*Cymbidium* spp.	Agglutinin	Antiviral
9.	*Dendrobium aduncum*	Aducin	Used as a tonic and in the treatment of fever
10.	*Dendrobium chrysanthum*	s-Dendrochrysine, Trans-Dendrochrysine, Hygrine, Erianin	Antipyretic, eyes-benefiting, Immunoregulatory purposes, skin diseases
11.	*Dendrobium monoliforme*	Dendromoniliside A, Dendromoniliside B, Dendromoniliside C, Moniliformin	Tonic and antipyretic, longevity and as an aphrodisiac, stomachic, and analgesic
12.	*Thunia alba*	Thunalbene	Treatment of dislocated bones
13.	*Vanda cristata*	Laburine acetate	Used as an expectorant and in bronchitis, cuts, wounds, boils, and dislocated bones
14.	*Vanda parviflora*	Parviflorin	Used to treat rheumatism and nervous system disorders, also as an anticancer and has antiviral properties
15.	*Vanilla planifolia*	Vanillyl methyl ether, Piperidinic acid	To treat hysteria, fever, and rheumatism to increase the energy production of the muscular system

biotechnological approaches and techniques become pivotal for the efficient propagation of orchids and their valuable secondary metabolites. In this context, knowledge of biosynthetic pathways is the key to secondary metabolites from different medicinal orchid species, which becomes imperative to enhance their production via *in vitro* techniques (Gantait et al., 2021).

The paste of pseudobulb of *Coelogyne stricta* has been used to cure headaches and fevers in traditional medicine systems.

It has been observed that in wild orchids and *in vitro* raised plantlets, protocorms are an essential source comprising various crucial bioactive components for the propagation of the orchids. The cytotoxic activity has also been mentioned for the first time in this chapter. Also, an effective conservation strategy is optimized for orchids through *in vitro* propagation techniques.

3.2 CONSERVATION OF MEDICINAL ORCHIDS

Orchids being illegally traded and over-exploited has led some species to the brink of being endangered or extinct, which has raised great concern about preserving these species. Several measures are being taken to conserve the orchids. The efforts are majorly concentrated on propagation to decrease harvest from wild populations, reacquaint plants in the wild populations, procure plants from void habitats, and add seeds to the global system of orchid seed banks. Obtaining superior seeds, i.e., mature, uncontaminated, and highly sprouting seeds, is vital for the successful conservation of orchids. Ex-situ propagation has not been extensively employed in temperate and tropical environments for native species conservation. Orchid seeds, also known as "dust seeds", are little, wind-dispersible seeds that are abundantly produced by each orchid bloom. A seed pod may contain as few as 20 or 50 seeds or as many as 4,000,000. The undifferentiated embryo, small reserves, and testa-covered air gaps observed in orchid seeds are all regarded to contribute to the extended periods of seed dissemination. The germination and storage potential of seeds is determined by species-specific traits such as the size of the embryo concerning the space within it, the presence of an outer shell enveloping the embryo, and the prevalence of chemical components and growth regulators. A symbiotic association with mycorrhiza is necessary for orchid seeds to germinate in the natural environment. Orchids are propagated through *in vitro* symbiotic germination (Pujasatria et al., 2020). Successful propagation, however, necessitates an in-depth understanding of the particular orchid-mycorrhizal interaction and the capacity to isolate and cultivate the fungus, which can take a significant amount of time and effort. High-quality pollinia for pollination also requires careful handling. To cryopreserve the germplasm of Indian orchid species, the first step is to devise methods of handling the various explants so that the materials can be safely transported from one location to another before long-term storage under viable conditions. Northeast India, a mega-diversity centre, comprises eight states, viz., Arunachal Pradesh, Assam, Manipur, Meghalaya, Mizoram, Nagaland, Sikkim and Tripura. It occupies 7.7% of India's total geographical area supporting 50% of the flora (ca. 8000 species), of which 31.58% (ca. 2526 species) are endemic. Conservation of plant germplasm is a need not only for food security but also for future sustainable use because the loss of germplasm diversity occurs due to man-made disasters such as urbanization, population explosion, deforestation, etc.

3.2.1 Importance of Orchids in Traditional Medicine

Orchids have also been used as medicine for a long time in many countries, such as China, as well as in some regions, such as Europe, America, Australia, and Africa. The ethnobotanical value of many orchids is also discussed in the Charka Samhita, an Indian medical classic written in Sanskrit by Charaka thousands of years ago. In the Middle Ages, people believed that the orchid plant came from sperm that fell to the ground among the grasses where animals gathered (Schweinfurth, 1959). Europeans believe that orchids are symbols of sexuality. Cypress root contains essential oil, tannin, sugars, starch, and other components. The root of this species is also used to treat stomach disorders and reduce joint pain. The medicine in orchids is important to reduce fever, increase white blood count, treat eye disease, treat fatigue and headaches, and, most importantly, prevent disease as an antidote. Approximately 40 *Dendrobium* species are used in Chinese medicine; five of them (*Dendrobium chrysanthum, D. fimbriatum, D. loddigesii, D. noboile* and *D. officinale*) are all listed in the Chinese pharmacopoeia. India has many medicinal and aromatic plants used in Ayurveda, homeopathy, allopathic and other traditions. Ayurveda and traditional Chinese medicine are important in discovering the benefits of this herb. In India, orchids have many medicinal uses in different traditional medicine systems. In the Ayurvedic system of medicine, a group of eight medicines called Ashtanga are used to prepare many rejuvenating preparations and medicines. Asthavarga is an important component of many classical Ayurvedic formulations, such as Chavan prasa (Singh and Duggal, 2009). Vanda has a long history of being used by local people for its anti-inflammatory properties. Vanda has anti-inflammatory properties against many types of cancer, including choriocarcinoma (cancer) and lung and stomach cancer (Ho and Chen, 2003). Vanilla is the essence of natural vanilla flavor. Fresh and dried stems of *Dendrobium* can be used to prepare medicines that act as aphrodisiacs and analgesics as long-term medicines (Uma Debi et al., 2009). Some species of orchids are used as antidotes to spider bites and to treat pain. Paste obtained from *Acampe praemorsa*, *Lusia zeylanica* leaves, and *Cymbidium aloifolium* roots can be used in the treatment of osteoporosis in humans (Behera, 2008). *Dendrobium*, an important Ayurvedic orchid, is reported to be the root of Jivanti (Kasera and Shukla, 2001). The oldest Chinese Materia medica lists 365 species of plants., including many orchids. Even today, *Dendrobium* and pseudobulbs are sold in the market under the name "Purrusharatna". Orchids, especially the *dendrobium* variety, have been used as medicinal plants in many countries. The uses of dried orchids include strengthening the immune system, improving vision, regaining strength after treatment, and making men stronger (Singh and Sandhu, 2005).

3.3 ORCHID CONSERVATION IN INDIA

3.3.1 Orchidarium and Orchid Sanctuary

Orchids had a special place in the life and culture of ancient Indians, who were directly or indirectly cultivated in the hermitages of great sages and shrines created by kings. Orchids in India are now considered protected plants. In recent years, many nature reserves have been established in various phytogeographic regions of India. Some states have begun to protect orchid habitats in protected areas specifically; for example, in Apangala, Lori Yangon in Karnataka, and Darjeeling district in West Bengal, the orchid has been designated as a conservation zone of the state Governments. Orchid sanctuaries have been established at Deorali and Singtam in Sikkim and Sessa in Arunachal Pradesh (Hegde, 1983, 1984). Protection, maintenance, and management of orchid habitats are the main motivations for orchid conservation projects. Regional Plant Resource Center (RPRC), Odisha, India, maintains living species and performs large-scale propagation under *in vitro* or *ex situ* culture.

3.3.2 *In Vitro* Conservation through Micropropagation

Tissue culture methods have been extensively employed for the past fifty years to conserve orchids in their natural habitat as well as for the quick and extensive propagation of orchids. Although conventional methods such as shoot division and trailing bulbs are commonly used to propagate orchids, the majority of orchids are sexually reproduced (Vij, 1995). Despite this, the growth of orchids is relatively slow, and only a few plants are produced even after a long period. Various protocols have been developed for the large-scale propagation of several orchid species through in vitro cultivation of various parts, including shoot tips, root tips, and nodes of various regions. *In vitro* cultivation for mass production of orchids is a difficult task. Therefore, large-scale propagation using traditional methods and tissue is an important strategy to protect the world population from commercial exploitation (Murashige and Skoog, 1962).

3.3.2.1 Embryo Culture

Since the development of the embryo sac is an event that occurs after pollination and fertilization is required to produce the seed, the young ovaries will not produce a viable plant. Additionally, the development stage of successful ovule germination varies from species to species, as eggs are ready for culture immediately after fertilization (Hossain et al., 2008). The symbiotic germination potential of seeds representing different developmental stages was tested in different threatened taxa of Indian orchids (Pant and Gurung, 2005).

3.3.2.2 Meristem Culture

Orchids are outbreeders and produce a large number of heterozygotes in their offspring. Thus, pure lines of the desired genotypes are favored in the cut flower industry, suggesting that

FIGURE 3.2 Handling and seed extraction of orchid species: a) Seeds of orchid in compound microscopic view, b) Collection of orchid fruit(pods) for seed extraction, c) Extract the seeds from a single fruit(pod) of an orchid.

propagation through embryo culture is a major disadvantage. The possibility of using excised shoot meristems for the regeneration of complete Cymbidium plants in vitro and the formulation (Figure 3.2), description, and publication of a procedure for this purpose, the meristem cultivation technique (shoot, auxiliary bud) opened new horizons in orchid micropropagation (Arditti and Ernst, 1993).

3.4 PROBLEMS AND PROSPECTS FOR THE CONSERVATION OF ORCHIDS

An in-depth knowledge of orchid reproductive biology is necessary for both conservation and management, as is a commitment to safeguarding their environment. It is necessary to prevent the eradication of species, particularly endemic species, by restoring threatened species to safe habitats and shielding humans from novel risks. Forest ecosystems were destroyed within a few years due to strong interaction between tribes and local traders. In 1975, the Convention on International Trade in Endangered Species (CITES) of Endangered Species of Wild Fauna and Flora proposed various conservation measures for orchids (Vij, 2001). Protection of natural habitats by establishing sanctuaries, biosphere reserves, and forest reserves; other proposed measures include the rescue of plants from damaged and endangered habitats and their cultures in orchidariums, botanical gardens, and another rescue areas, and other proposed measures include in vivo or in vitro propagation of endangered plants and their replanting in well-protected habitats.

3.5 SOME IMPORTANT RARE MEDICINAL ORCHID SPECIES

As mentioned, *in vitro* propagation is the only remaining tool for conserving some rare and endangered species:

(I) ***Coelogyne nitida***: It is one of the temperate species restricted to only the Himalayan region of India and Southeast Asia. *Coelogyne* is a shining, evergreen, epiphytic orchid with dark green leaves, which can be up to a foot long. *Coelogyne* occurs on forest trees such as rhododendron, Quercus, and oaks in the Himalayas from Uttarakhand to Bhutan and Burma at an altitude of 1,500–2,500 m above sea level.

Medicinal use:

- Patients having severe headaches are said to be relieved from this plant whole paste.
- It is said that the pseudo bulbs have aphrodisiac properties; even the juice of pseudo bulbs is applied in case of wound healing.

(II) ***Cymbidium mastersii*:** This species belongs to an epiphytic orchid with a cosmopolitan distribution in Northeast India. Since its valuable, big, attractable flowers and highly aromatic feature have made it nearly extinct in natural habitats, it is always in great demand in the floriculture industry and can only be grown in in vitro labs since its natural habitats have been destroyed.

Medicinal use:

- The whole plant paste is used for healing wounds, bone fractures, and dislocated bones.
- The concentrated decoction is used by tribes traditionally in severe coughs and colds.
- The plant's fresh juice is used to stop both internal and external bleeding.

(III) ***Cymbidium*** **species:** Several Cymbidium species, namely *C. canaliculatum* R. Br, *C. madidum* Lindley, *C. eburneum* Lindl., *C. aloifolium* (L.) Sw., *C. devonianium* Lindl. ex.

Paxton, *C. iridioides* D. Don, *C. giganteum* Wall. Ex Lindl., and *C. sinense* Willd., are used as medicinal plants in the traditional medicine of many Asian countries (Pant and Rinchen, 2012). Thus far, a wide range of substances, including phenols, alkaloids, phenanthrenes, stilbenoid derivatives, and steroids, have been isolated and identified, and their molecular structures have been identified by a variety of spectroscopic techniques. Leaf extract is used to treat fevers and ulcers. It is also used as an emetic, tonic, and laxative; it can heal burns, wounds, and earaches. Crushed plant extracts with ginger are given to treat eye weakness, dizziness, chronic diseases, and paralysis. Recent reports indicate that *Cymbidium* root extracts have high antimicrobial activity against Staphylococcus aureus, and stem extracts contain phenolic compounds that exhibit high antioxidant activity and cellular cytotoxicity.

(IV) ***Thunia* alba:** This species is known as white Thunia, which grows and creeps on the ground. It makes straight, biennial leafy canes ending as white decorative *Coelogyne-like* flowers bearing yellow and orange distinct, marked lips. White Thunia commonly grows up to an altitude of 2000 MSL in North East regions of India, North Burma, Thailand, and South China. It's used as the same *Cymbidium iridioides* (Awasthi and Chauhan, 2020).

(V) ***Dendrobium* hybrid:** There are more than 1,200 *dendrobium* species that are native to the land of tropical Asia and the Pacific region. *Dendrobium* is found to be grown on crown parts of trees; hence, they are also known as jewels of trees. *Dendrobiums* are cosmopolitan and have natural habitats from tropical to lowlands all also at high altitudes up to 3,000 meters above sea level.

Medicinal use:

- Traditionally, the fresh paste is applied to treat burns and scalds, skin itching caused by a red rash, and eczema.
- Fresh pseudobulbs are ground with water to yield a juice or fresh pseudobulbs are pounded and externally used to treat scalds. A decoction of dry pseudobulbs (15 g) is taken orally to treat skin itching caused by red rash.
- The firmly ground paste is used for wound healing and dislocated bones.
- The fine paste of pseudo bulbs is used for bowels and acne.

Source of orchids: The viable seeds of different rare medicinal orchids were collected from naturally wild pollinated plants, which are growing mostly in the Northeast region of India. Collected orchid pods have been surface sterilized under invitro conditions, and seeds have been extracted carefully under laminar air-flow or aseptic conditions. Seeds have been grown in different mediums and recorded data through viability and germination.

3.6 POST-HARVEST HANDLING OF ORCHIDS

Post-harvest parameters include varietal differences, the amount of light, flower sugar level, temperature, nutrition, and water loss; harvest variables like time, method, and harvest stage; and post-harvest variables like ethylene production, pre-chilling, water purity, pulsing, employing preservatives, bud opening, packaging, and storage all have an impact on how long orchid cut flowers last after harvest.

3.6.1 Pre-Harvest Factors

It is believed that a third of the post-harvest longevity of orchid flowers is dependent upon the pre-harvest handling of the flowers.

TABLE 3.2
Vase Life of Orchids Due to Species and Varietal Differences

Name of Species/Hybrids	Vase Life (Days)
Aerides, Dendrobium nobile, Renanthera imschootiana, Aerides multiflorum, Cymbidium iridioides	28–56
Paphiopedilum hirtussimum, P. wardianum	56
Phaius tankervillae	28–42
Vanda coerulea, Vanda teres, Zygopetalum intermedium	14–21
Cymbidium hybrids	20–55
Dendrobium hybrids	14–21
Vanda, Mokara hybrids	14–30
Vanda, Mokara hybrids	10–20
Phalaenopsis hybrids	25–30
Aranda	18–28

3.6.1.1 Varietal Differences

Due to variances in water intake, fresh weight, flower diameter, stem lignification, vase life, and aging, varying cut flower variabilities have been documented (Table 3.2).

Among the different species, vase life ranges are as follows: *Phaius tankervillaea* (24 days), *Zygopetalum intermedium* (18 days), *Aerides multiflorum, A. odoratum, Cymbidium iridioides, Dendrobium nobile* and *Renanthera imschootiana* (28–56 days), *Paphitisopedilum hirsuwardianum* (56 days), and *Vanda coerulea, Vanda teres* (28–42 days).

3.6.2 Harvesting Factors

3.6.2.1 Harvest Time of Flowers

Moderate temperatures are ideal for flower harvesting since high temperatures accelerate respiration and cause excessive water loss. It is best to collect flowers early in the morning or late at night. Because of their elevated sugar content and night sweats, the flowers are still swollen in the early morning hours. Similar to this, if cut in the afternoon, flowering stems hold on to more stored carbs and have a longer vase life.

3.6.2.2 Method of Harvest

The flowering stalk ought to never be removed from the parent plant without the use of sharp instruments and pruning shears. When harvesting, the stem, especially hardwood stems, should not be crushed, and the cut should be made at an oblique angle. After harvesting, the spikes need to be submerged in a vessel of water right away. Harvest stage: For commercial orchids, the completely opened and fully developed stage of the harvested flower, as well as the length and quantity of spike, are the ideal conditions for harvesting.

3.6.3 Post-Harvest Factors

3.6.3.1 Temperature

At warmer temperatures, the rate of senescence and bud opening fastens. Flowers release less ethylene, and respiration slows down at lower temperatures. For buds to completely expand, flowers must be plucked at an immature stage. For *Cymbidium* and *Paphiopedilum*, buds are kept at temperatures as low as 0.5 to 4.0 °C; for *Dendrobium,* 5.0 to 7.0 °C; and for *Cattleya,* 7.0–10.0°C.

3.6.3.2 Light

Extended storage or long-distance transportation of cut flowers necessitates light. Similar to this, tightly wound flowers into buds require intense light to unfold. Extended dark storage of flowers, like carnations, does not compromise their quality. In their lighting businesses, florists should keep the light intensity between 2,000 to 3,000 lux for 12 to 24 hours.

3.6.3.3 Humidity

For optimal preservation of the turgidity of cut flowers, they should be kept at 90–95% relative humidity. Wilt symptoms begin when flowers lose 10–15% of their fresh weight. As relative humidity increases, the leaf transpiration rate decreases.

3.6.3.4 Water Quality

The hardness concentration of phytotoxic substances, pH and EC values, and illness that clog vessels and shorten the vase longevity of cut flowers are all considered indicators of water quality. Salt water reduces the life of a vase of cut flowers. For cut gladiolus, flower life is reduced when the concentration of salts in the water reaches 700 ppm. The lifespan of the flowers is short, and 200 ppm poses a problem when cutting roses, chrysanthemums, and carnations. More than 200 ppm of salt reduces vase life by half a day for every 100-gram increase in salinity. Alkaline ions such as calcium and magnesium present in hard water are less harmful to flowers than soft water containing sodium ions. Fluoride is extremely toxic to most cut flowers, damaging freesias gladiolus and gerbera at 1 ppm, and chrysanthemum and roses at 5 ppm. Flowers such as lilac, cymbidium, and daffodil are resistant to fluoride ions. Passing tap water or quality water through a deionizer will extend the life of the vial. A control solution with a pH of 3.0–5.0 is best to extend the vase life of cut flowers.

3.6.3.5 Ethylene

In climacteric flowers, ethylene is crucial to the control and synchronization of bloom senescence. In buds and early blooms, this hormone is not consistently produced. At the ripening, senescence, and flower opening stages, there is an increase in ethylene evolution. Ethylene production then levels out and stays at the same level. The leaves autocatalyze more ethylene that would otherwise cause petal rolling because, essentially, ethylene is first created in the pistil and then evolves to the skin where it produces and stimulates the expression of ACC synthase, ACC oxidase, and cysteine protease genes and flower wilting (Figure 3.3). Pistils produce more ethylene in flowers; this can be instigated by factors such as ABA or IAA, demostrates the importance of pistils in the regulation of floral ethylene

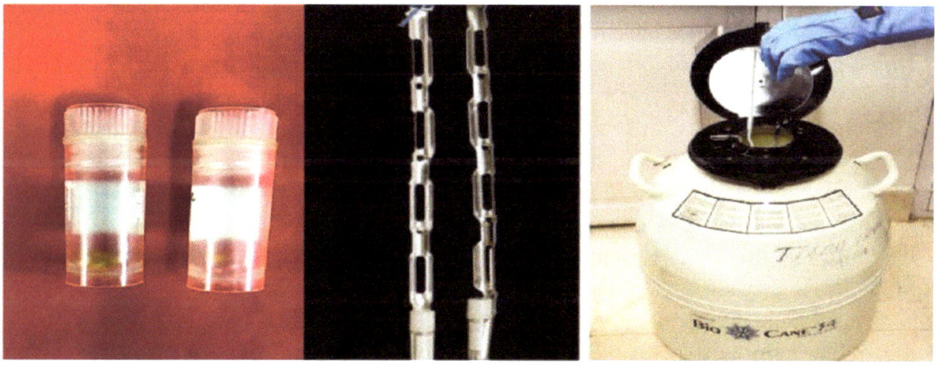

FIGURE 3.3 A complete procedure of cryo-preservation of pollens: a) collection and store of pollen in cryovials, b) cryovials setup in cryostics, c) cryovials(with pollen) in liquid nitrogen for cryopreservation.

during natural and pollination-induced senescence, with emasculation accelerating ethylene release. Ethylene affects a wide variety of flowers, causing symptoms like carnation drowsiness, five-clawed flower petal curling, orchid sepal tips drooping, and anthocyanin production. The flowers of orchids are extremely ethylene-sensitive. Pollination, mechanical injury, and herbivore damage all result in high levels of ethylene production. Compared to untreated flowers, decapacitated and emasculated blooms release more ethylene. Sometimes, the forced unfolding of flower buds can cause their lifespan to be shortened. Pests and diseases like fungi, bacteria, and insects can produce too much ethylene, which affects the quality of cut flowers. Microbes damage by clogging xylem vessels and producing ethylene and toxins from pectin degradation compounds. Among bacterial species, *Alcaligenes, Pseudomonas, Enterobacter, Erwinia, Bacillus, Corynaebacteria, Aeromonas, Acetinobacter,* and *Flavobacterium* are commonly found in the vase. Some species of fungi, such as *Botryris cinerea, Fusarium oxysporum, Mucor, Penicillium* spp., *Rhizopus, Aspergillus* spp., *Alternaria alternata*, and *Acremonium strictum,* can be responsible for premature senescence of flowers, wilting and rotting of potted plants.

3.7 CONCLUSION AND FUTURE REMARKS

Numerous orchid species have declined, and information about them has been lost as a result of the ongoing exploitation of various orchid species in the wild, the significant loss of the orchid environment, and the deterioration of customary rules managing natural resources. Orchids are highly responsive to changes in their surroundings. For this reason, it is crucial to preserve these priceless orchids and traditional knowledge to advance pharmacological research into their potential medical benefits.

The study confirmed that careful rapid transport of pods, seeds, and pollinia of rare medicinal orchid species is essential. The explants showed good viability on the observation receipt, with no transportation losses.

The following lessons were learned based on:

- Immature seeds do not tolerate drying and, hence, cannot be processed for seed cryo-banking.
- Herbarium specimens must also be kept for correct information. However, it is generally not possible due to detrimental effects, and hence, photos of flowers, pollens, and pods need to be kept systematically.
- Capsule maturation markers have to be recorded to provide data for identifying suitable harvest times. It was better to sterilize the pods and use the extracted seeds under sterile conditions than to sterilize the seeds.
- Nutrient media- Murashige and Skoog, Knudson "C" orchid, and New Dogashima medium were the best for seed germination of different orchid species.
- Easy handling methods have been proposed for sensitive tissues such as pods, seeds, shoot tips, and PLB of orchids for conservation, good growth, and recovery.

When it comes to flowers and medicinal plants, there is a high demand for ornamental and medicinal plants in the world market. Orchids have both ornamental and medicinal value and are highly sought after in the global market. Research on the antimicrobial properties of orchids is a major task for conservationists in the field. Endemic and rare orchids with ethnomedicinal properties should be pursued to explore their uses for human welfare. Numerous nations across the globe are now witnessing financial success in the creation of novel hybrid orchids and their commercial cultivation. For the sake of commercialization and conservation, appropriate strategies for the large-scale multiplication of new hybrids, genetically modified orchids, and rare, threatened, and endangered orchids should be established. The scientific community believes that the primary threat to the preservation of these species in their natural habitat is the loss of biodiversity. Although India has laws to protect flora and fauna, a lack of awareness among people can destroy the forest ecosystem. Additionally, orchids are mostly terrestrial and epiphytic, meaning that large die-offs and forest fires can damage orchid habitats. In situ conservation is part of strategies to conserve orchid species, but there is an urgent need to preserve translocations in their natural habitat.

3.7.1 Author Contributions

M.A. contributed to the design, wrote parts of the manuscript, and prepared materials. A.C. collected data. S.M. revised drafts and final preparation of the manuscript. All authors read and approved the final version of the manuscript.

REFERENCES

Arditti, J. and Ernst, R. 1993. Micropropagation of orchids. John Wiley and Sons, New York.

Awasthi, M. and Chauhan, J. 2020. Conservation of Some Selected Rare Medicinal Orchid Species through Post-Harvest Handling of Different Seeds, Pods and Pollinia. *Ecology Environment and Conservation*, **26**:20–25.

Behera, D. 2008. *Ethnobotanical studies of some selected orchids and screening of their antimicrobial potential*. M. Phil. Thesis, North Orrisa University, Orrisa, India.

Gantait, S.; Das, A.; Mitra, M. and Chen, J.T. 2021. Secondary Metabolites in Orchids: Biosynthesis, Medicinal Uses, and Biotechnology. *South African Journal of Botany*, **139**:338–351.

Hegde, S.N. 1983. Orchid Sanctuary, Sessa: An Effect Towards Habitat Conservation, Seminar on Conservation in Developing Countries-Problems and Prospects, Bombay.

Hegde, S.N. 1984. Role of Orchidarium and Orchid Sanctuaries in Conservation Research and Development of Orchidology in Arunachal Pradesh. In: Proceeding of the 5th Asian Orchid Congress, Singapore.

Ho, C.K. and Chen, C.C. 2003. Moscatilin from the Orchid *Dendrobium loddigesi* is a Potential Anticancer Agent. *Cancer Investigation*, **21**:729–736.

Hossain, M.M.; Sharma, M. and Pathak, P. 2008. In Vitro Mass Propagation of an Economically Important Orchid *Cymbidium aloifolium* (L.) *Sw. Journal of Orchid Society*, **22**:91–95.

Kasera, P.K. and Shukla, J.K. 2001. Biomedical Properties and Cultivation of Leptadaenia Reticulata and Endangered Plant of the Thar Dessert, India. *Current Science*, **84**(7):877–880.

Kaur, A.; Verma, J.; Yadav, V.G.; Pawar, S.V. and Sembi, J.K. 2023. Exploring the Potential of In Vitro Cultures as an Aid to the Production of Secondary Metabolites in Medicinal Orchids. *Advances in Orchid Biology, Biotechnology and Omics*, **1:**163–185.

Mudoi, K.D.; Borah, P.; Gorh, D.; Gupta, T.; Sarmah, P.; Bhattacharjee, S.; Roy, P. and Saikia, S.P. 2023. Biotechnological Interventions and Societal Impacts of Some Medicinal Orchids. *Advances in Orchid Biology, Biotechnology and Omics*, **1:**59–144.

Murashige, T. and Skoog, F. 1962. A Revised Medium for Rapid Growth and Bioassays with Tobacco Tissue Cultures. *Physiology Plantarum*, **15**:473–497.

Pant, B. and Gurung, R. 2005. In-Vitro Seed Germination and Seedling Development in Aerides Odorata Lour. *The Journal of the Orchid Society of India*, **19**(1–2):51–55.

Pant, B. and Swar, S. 2013. Micropropagation of Cymbidium Iridioides. *Nepal Journal Science and Technology*, **12**:91–96.

Pant, S. and Rinchen, T. 2012. *Dactylorhiza hatagirea*, a High-Value Medicinal Orchid. *Journal of Medicinal Plants Research*, **6**:3522–3524.

Pujasatria, G.C.; Miura, C. and Kaminaka. H. 2020. In Vitro Symbiotic Germination: A Revitalized Heuristic Approach for Orchid Species Conservation. *Plants*, **9**:1742.

Schweinfurth, C. 1959. Classification of orchid. In: Wither, C.L. (Ed.) Orchids scientific survey, Ronald Press Co., New York, pp. 15–44.

Singh, A. and Duggal, S. 2009. Medicinal Orchids: An Overview. *Ethnobotanical Leaflets*, **13**:351-363.

Singh, A.P. and Sandhu, A.S. 2005. A dictionary of medicinal plants. Singhal S, Sundeep Publishers, New Delhi.

Uma Debi, P.; Selvi, S.; Devipriya, D. and Murugan, S. 2009. Antitumor and Antimicrobial Activities and Inhibition of Invitro Lipid Peroxidation by *Dendribium nobile. Journal of Biotechnology*, **8**(10):2289–2293.

Vij, S.P. 1995. Orchid Genetic Diversity in India. Conservation and Commercialization. In: Proceeding of the 5th Asia Pacific Orchid Conference and Show, Fukuoka, 20–39.

Vij, S.P. 2001. Orchidology in India. Current Status. In Hegde, S.N. (Ed.) Orchids conservation, culture, farming and trade. Orchid Society of Arunachal Himalayan Publishers, Delhi, Itanagar.

White, K.J. and Sharma, B. 2000. Wild orchids in Nepal: The guide to the Himalayan orchids of the Tribhuvan Rajpath and Chitwan Jungle, White Lotus Press, Bangkok.

4 Traditional Medicinal Uses of Orchids

B.D. Sharma

4.1 INTRODUCTION

Different species of orchids distributed globally are not just valuable ornamental plants but are also used in traditional herbal medicines. Orchids are a large group of plants with over 25,000 species and have several varieties, including hybrids. In India alone, there are about 2,500 species belonging to 167 genera (Singh et al., 2012). The first medicinal use of orchid species can be traced back to about 4500 BC during the Harappan Civilization when they were discovered by the great Vedic period physicians, the Ashwani Kumars. The four orchid species *Habanera intermedia* D. Don (Rddhi), *Platanthera edgeworthii* (Hook.f. ex.Collett) Vrddhi, *Crepidium acuminatum* (D.Don) szlach (Syn. Malaxis acuminata D.Don; *Microstylis wallichii*) Jeevak, and *Malaxis muscifera* (Lindl.) Rshlbhaka, part of the Astavarga group, has been used to make the world's first "Nutraceutical, chayavanprash".

Traditional knowledge of medicinal uses of plants is the most valuable treasure for achieving good health. "Ethnobotany", or precisely, "Ethnomedicine" is the scientific discipline of the study of the interaction of human beings with plants. It has many applications in several fields, addressing current global concerns such as human health care, food security, nutrition, climate change, and biological diversity conservation (Aswandi and Kholibrina, 2021). The documentation and maintenance of traditional knowledge of medicine that have been empirically developed over a long time will pave the way forward for modern research (Sharifi-Rad et al., 2022).

Plants have been the main ingredient for treating various human ailments since time immemorial, especially in remote areas inhabited by tribal or other weaker sections of society. Indigenous practitioners and traditional health care practitioners throughout the world have gained rich knowledge of different plants. These people also have expertise in the identification, collection, and application of this knowledge for health benefits. The idea of utilization of plants for local remedies is to provide a cheaper, easier, and more sustainable alternative to synthetic drugs. It is proven that orchids have many active components that have the potential to cure serious ailments to a high level of satisfaction. Traditional medical knowledge is often passed on orally from generation to generation. Traditional healing practitioners and remedies made from plants play an important role in maintaining the health of millions of people even today. Therefore, the study of ethnomedicines is useful in documenting, analyzing, and communicating knowledge of a plant and its interaction with human society. Orchids have an important position in traditional medicine and, hence, need to be explored more for their application in modern pharmaceuticals (Singh and Duggal, 2009).

4.2 TRADITIONAL USES

The medicinal uses of orchids are recorded in the Harappan Civilization and Vedic period. In Atharva Veda, Panini Astadhavi, and Vedic literature, there is no mention of these four orchids, including the Astavarga plants. However, Dhanwantiri Nighantu made the first mention about 3500 BC (Sharma, 2005). The Chinese were the first to publish a book on about 20 orchid species and their cultivation in 1233 (Varma and Ingalhalli, 2022). In the Indian Materia Medica (Nighantus-Dravayaguna Vijiana), besides the mention of four temperate orchid plants, some other important orchids which

DOI: 10.1201/b23375-4

are largely used in Ayurvedic drug formulations, including Rasna, Nakuli, *Acampe papillosa* Lindl., *Acampe carinata* Griff, *Acampe praemorasa* Roxb; Salabmisri, Pranada, Varakanda *Eulophia dabia* (D.Don) Hochr. (= *Eulophia campestris* wall); Rasna, *Vanda cristata* Lindle.; Rasna, *Vanda tessellata* Roxb. Hook. ex, Don; Salampanja, Hathajari, Munjataka *Dactylorhiza hatagirea* (D.Don) Soo; Jivanti, Svarana Jivanti, Yashasvani, Jiva Bhadra, *Desmotriohum fimbriatum* Bl., *Dendrobuim Macraei* Auct.); Amarkanda Malakanda, *Eulophia nuda* Lindle. (Syn.*E. dabia*). The four orchids from the Astavarga group are said to be Jeevaniya (Vitality promoting), Brhaneaya (promote the formation of mansa dhatu (body flesh) and vayesthapana (gerontologic).

4.3 CLASSIFICATION DEPENDING UPON AVAILABILITY AND USES

About 165 medicinal orchids found in India can be grouped into three categories.

1) Highly valuable rare temperate Himalayan Orchids (high demand in Ayurvedic medicines),
2) Orchids generally used in Ayurvedic formulations, and
3) Medicinal orchids used as traditional medicines in various parts of India.

4.3.1 High-Value Rare Temperate Orchids

As mentioned earlier, four medicinally highly demanded orchids were known. However, it is recent that their botanical identity was first documented in the book, "Vitality Strengthening Astavaraga Plants" (Sharma and Balkrishan, 2005). They (four orchids), along with others (two from Asparagaceae: one from Liliaceae and one from Zingiberaceae), constitute the well-known Astavarga constituting the "Chavayanprash". It is the first ever known and the most ancient concept of "functional food" in the world. Chayavanprash was named after "Vedic period Rshi Chayavan", who concluded the curing effects of all eight herbs on various respiratory diseases, including tuberculosis. The medicinal properties of various orchids described in the Indian Materia Medica (Nighantus) are mentioned in what follows. However, in Shaligram Nighantu it was mentioned that the availability Astavarga plants was rare to find. The description of four plants given here is from the "Dhanwantary Nighantu" (oldest Nighantu) and same was repeated in later Nighantus viz Bhavapravash etc. (Singh, 2022).

1) ***Crepidium acuminata* (D.Don) szlach. Syn. Jeevaka:** Pseudotuber-bearing terrestrial orchid, which is "Tha Triman" (sweet) and cooling (Sheeta veerya). It pacifies vata, pitta, and rakta doshas, is a cure for tuberculosis (Kshaya) and fever, and increases Kapha dosha and seminal fluid (sukra). It is beneficial as an emollient, aphrodisiac, rejuvenating, tonic, and is a treatment for cough, epilepsy, insanity, emaciation, burning all over the body, and general debility (Sharma, 2021).
2) ***Habenaria intermedia* D.Don (Rddhi):** It is tuber bearing temperate orchid. The tuber is sweet in taste and refrigerant (Sheeta Veerya). It is a treatment for tuberculosis, vitiated vata, pitta rakta doshas, and a cure for tuberculosis (ksyaya), fever, and increases kapha doha and seminal fluid (shukra dhatu). It is useful in heart problems and fevers. It is also considered intellect promoting, an aphrodisiac, rejuvenating, tonic, and is a cure for emaciation, epilepsy, cough, skin diseases, hematemesis (vomiting of blood), and general weakness.
3) ***Malaxis muscifera* (Lindl.) Kuntze (=*Microstylis muscifera* Lindl.) (Rshabhaka):** It is a rare species because it grows in grasslands or pasture lands. The pseudobulbs are an efficacious treatment for disordered Vata, pitta, and rakta doshas (internal bleeding problems) and are useful in treating phthisis, burning sensations all over the body, and fever. The plant is used to increase kapha dosha and shukra dhatu (seminal fluid), and is styptic, aphrodisiac, tonic, and is an effective remedy for dysentery, body thinness, seminal weakness, and general body weakness (Sharma, 2021).

4) ***Platanthera edgeworthii* (Hook. and. ex. Collett) RK Gupta syn. *Habenaria edgeworthii* (Hook. f. ex Collett) Vrddhi**: This orchid is terrestrial in habit and grows in association with tall grasses. It has become rare because it is cut before seed maturation along with grass and is used to feed domestic animals. Sometimes, if there is fire, tubers are roasted and rendered incapable of germination. The tubers are sweet (madhura) and Sheeta veerya (refrigerant), which are known to treat kapha dosha and increase seminal fluid (shukra), pacify vata and pitta doshas, and act as cure for emaciation and tuberculosis and are included in Hritikyaadi varga (Bharvaprkasha, Verse 10). Dhanawanteri Nighantu (142–143) includes it in Guduchiyadi varga and mentions that its tuber is sweet and it is a remedy for treating aggravated pitta, rakta (blood) doshas, heals burning sensation, phthisis, and fever due to vitiation of vata dosha. Raja Nighatau and Kaiyadeva Nighantu also describe similar medicinal properties for the herb. In some other Nighantus, it is mentioned that it is a cure for sterility, dysentery, and general body weakness.

4.3.2 Orchids as Ayurvedic Herbal Medicines

Traditionally, Ayurveda practitioners or healers have been predominantly using herbs that are abundantly found in the country's rural inhabitations. Orchids have also been used widely in several Ayurvedic formulations and are mentioned in various Nighantus (Materia Medica). Important genera among them are mentioned in what follows.

1) ***Dactylorrhiza hatagirea* (D.Don) Soo (=*Orchis latifolia* L.):** It is popularly known as salampanja or Hathajari because its tubers resemble hands with fingers or look like folded hands. The salamapanja contains high amounts of mucilage, sugar, and albumen, and is rich in potash and calcium with traces of volatile oils, too. It has use as an expectorant, farinaceous food, nervine tonic, and aphrodisiac. The pseudobulbs are beneficial in dysentery, diarrhea, paralytic affections, and chronic fevers as well as valuable for the general debility of women after delivery (Sharma, 2021). In Gujrat, it is used to treat burns and wounds and to control bleeding (Varma and Ingalhalli, 2022).
2) ***Desmotrichum fimbriatum* Bl. (=*Dendrobium plicatle* Lindle.):** It is known as "Jivanti or Svrna Jivanti, saka Shrestha, Yashashwasni, Jiva Bhadra" in Ayurvedic Nighantus. Tubers are sweet and have an aroma. They are cooling, alterative, expectorant aphrodisiac. Tubers are useful for treating vitiated Vata, pitta, and kapha doshas, throat troubles, biliousness, diseases of eyes and blood, poisonous affections, tuberculosis, and asthma.
3) ***Cymbidium ebumeum* "Jivanti":** Leaves and seeds are used in medicines. Leaf juice mixed with oil is used in ptosis (paralysis of the upper eyelid), facial paralysis, headaches, and for treating solid tumors. Seed powder is used to heal ulcers due to diabetes (Panda and Mandal, 2013).
4) ***Eulophia dabia* (D.Don) Hochr. (=*E. camperstris* Wall. ex Lindley):** The orchid is known as "Salam misri or Pranda" in Ayurveda. It is an endangered species because its rhizome is of medicinal use. The rhizome is a valuable tonic and is used as an aphrodisiac. In Ayurveda, it is used to treat purulent cough, stomatitis, and heart ailments. It is said to be refrigerant, fattening (flesh forming), expectorant, and a treatment for heart ailments. It is demulcent anthelmintic and is a beneficial remedy for after childbirth weakness, tiredness due to overexertion, and diseases due to excessive indulgence in sexual activities (Sharma, 2021). It is mentioned as "Hattipaila" in Nepali, and its tubers are used as an appetizer, aphrodisiac, alterative, and it is useful for heart-related disorders (Varma and Ingalhalli, 2022).
5) ***Eulophia nuda* Lindle. (Amarkanda, Mankanda, Malakanda):** The tuber is known to generate heat in the body (increases pitta dosha), an appetizer, and is useful for treating worm infection, scrofula, tuberculosis of glands of the neck, tumors, and bronchitis caused by vitiated vata. The herb contains nudol-a phenathrene as an active component.

6) ***Eulophia spectabilis*** **(Dennst.) Suresh, Amarkanda:** Tuber and leaves of the plant are used as medicine. Its tuber is considered an appetizer, aphrodisiac, and blood purifier. It has been used as a remedy for tuberculosis of neck glands, tumors, and bronchitis. Decoction of the tuber is used as a treatment for roundworms and pinworm infection in children. Paste made from the leaves treats skin diseases (Tsering et al., 2017).
7) ***Acampe carinata*** **(Griff) Panigrahi:** It is known as "Rasna" in Ayurveda and is used in disorders of aggravated vata dosha. The root is used in acute conditions of rheumatism, sciatica, neuralgia, uterine diseases, and secondary syphilis. Root paste is used externally to treat snake bites and scorpion bites. Leaf decoction is given orally to relieve chest pain, stomach troubles, and acidity (Tsering et al., 2017).
8) ***Acampe papillosa*** **(Lindl.) Hochr:** This is also known as "Rasna" and is used as a substitute. The Naga tribe uses it for treating burns (Nongdam, 2014). It is a good remedy for rheumatic pains, sciatica, neuralgia, and urinary complaints. Paste from roots is used externally to treat snake bites and scorpion stings. A mixture of garlic with its leaf as a paste is a cure for acidity and other digestive disorders (Varma and Ingalhalli, 2022).

4.3.3 Medicinal Orchids as Ethnomedicines in India

In many parts of the world, orchids have long been used as traditional or ethnomedicines (Bulpitt et al., 2007). The first use of four orchids as ingredients of "Chayavanprash" about 5,500 years ago was the most ancient. However, the Chinese were the first to cultivate and describe orchids for medicinal use (Jalal et al., 2010). In Sowa-Rigpa, the traditional system of Tibetan medicine, and Ayurveda several species of orchids are mentioned for medicinal use. Other Asian countries, Africa, America, and Europe have a long history of the use of orchids as an ethnomedicine. Various orchid species used as ethnomedicines are described:

1) ***Acampe wightiana*** **Lindl. (syn. *Saccolabium praemorsum* Hook.f.):** A tonic made from the plant is used for treating rheumatic pain. Paste from leaves, heals bone fractures. Also used as a cure for typhoid fever due to its activity against *Salmonella typhosa* (Sharma, 2021).
2) ***Aerids crispum*** **Lindl.:** Used as a cure for earache. The whole plant is crushed and boiled in neem oil and then filtered. 2–3 drops of the oil are dropped in each ear. The pain-relieving nature is because of the presence of phenathropyran and aeridin (Sharma, 2021).
3) ***Aerides multiflorum,*** **Roxb. (=A. affine Lindl.):** Plant paste is a remedy for treating cuts and wounds externally (Varma and Ingalhalli, 2022). It has antibacterial activity against bacteria causing pneumonia and hence requires further investigation to develop an effective cure (Sharma, 2021).
4) ***Aerides odoratum*** **Lour.:** Fruit powder is used to heal wounds. Juice from leaves is a remedy for treating boils in the ear and nose. It is also a treatment for tuberculosis (Juice of fresh leaves consumed twice a day for 20 days). It is also used to treat joint pains and swellings. Paste of leaves mixed with neem bark, if consumed, cures joint pain and swelling (Yonzone et al., 2012).
5) ***Agrostophyllum bravipes*** **Ridly:** Powder from the tuber is a remedy for inflammation, wounds, and skin ailments (Panda and Mandal, 2013).
6) ***Agrostophyllum callosum*** **Rehb.f.:** Tuber powder is used to treat diabetes, inflammation, and skin ailments (Panda and Mandal, 2013). Tuber is a source of triterpenoids (Sharma, 2021).
7) ***Anoectochilus roxburghii*** **(Wall.) Lindl.:** It is a terrestrial orchid. The whole plant is taken orally to treat tuberculosis (Pant, 2013). Plant decoction is a cure for fever, lung diseases, and hypertension. Plant paste is used externally to treat snake bites (Nongdam, 2014).

8) ***Anthogonium gracile* Lindl.:** The roots contain a mucilaginous substance, and hence, it is used as "lep" for joint pain in Assam and other NE states of India (Yonzone et al., 2013).
9) ***Arundina graminifolia* (D.Don) Hochr.:** The whole plant, along with pseudobulbs, is crushed, boiled, and filtered to form extract. It is then consumed orally to treat diabetes, tumors, hepatitis, and hyperliposis. Tuber is used to treat body aches, wounds, fractures, and boils. Pseudobulbs boiled in coconut oil is a cure for cracked hands and feet (Panda and Mandal, 2013).
10) ***Appendiculata cornuta* Blume:** Stem juice is taken to treat fever.
11) ***Brachycortis obcordata* (Lindl.) summerh.:** A terrestrial orchid of eastern Himalaya. The root is a remedy for dysentery. Decoction is used as a tonic (Pant, 2013).
12) ***Bulbophyllum careyanum* (Hook.f.) Spreng:** Paste of leaves and pseudobulb treats burns. Powder of leaves is taken for inducing abortion before birth. After delivery, the plant is consumed for the recovery.
13) ***Bulbophyllum leopardinum* (Wall.) Lindl:** The juice or paste of the whole plant is a remedy for burns.
14) ***Bulbophyllum neilgherrense* Wight**: Decoction of pseudobulbs is used as a tonic and to provide a youthful look.
15) ***Bulbophyllum odoratissimum* (Sm.) Lindl:** Whole plant is given as infusion or decoction to treat tuberculosis, chronic inflammations, and bone fractures (Gutiérrez, 2010; Pant, 2013).
16) ***Bulbophyllum repens* Griff.:** The root is boiled and consumed orally for treating phlegm and as a cure for dry throat. It is also used to improve appetite and digestion (Aswandi and Kholibrina, 2021).
17) ***Bulbophyllum umbellatum* Lindl.:** The orchid is used as an analgesic and is anti-inflammatory.
18) ***Calanthe plantaginea* Lindl.:** Powder of dried rhizome is taken orally as a tonic and an aphrodisiac.
19) ***Calanthe puberula* Lindl.:** Powder of dried rhizome is taken as a tonic.
20) **Calanthe sylvatica (Thour.) Lindl.:** Flower juice is used to stop nasal bleeding. Leaves are used as a cure for colds and coughs (Panda and Mandal, 2013).
21) ***Calanthe triplicata* (L.) Fritsch.:** Roots chewed as a remedy for diarrhea. Flowers treat the pain of dental caries. Pseudobulbs are used as a masticatory for treating gastrointestinal problems (Yonzone et al., 2013).
22) ***Calanthe tricarinata* Lindl.:** Leaf paste is applied to sores and eczema as a cure. Leaves and pseudobulbs are made paste and taken as an aphrodisiac.
23) ***Cephalanthera longifolia* (L.) Fritsch.:** Powder of root and rhizome is used as a remedy for impotency and to increase body vigor and vitality.
24) ***Cleisostoma williamsonii* (Rechb.f.) Garay:** The whole plant paste is used as an astringent, used in skin care products to remove oil, tighten skin, etc. Fresh leaves and stems are used to treat bone fractures.
25) ***Coelogyne corymbosa* Lindl.:** Paste of pseudobulbs is a treatment for burns, wounds, cuts, etc.
26) ***Coelogyne cristata* Lindl.:** Pseudobulbs treat constipation and are used as an aphrodisiac. Juice of pseudobulb is useful to treat boils and wounds. A gum extracted from bulbs is beneficial for treating sores of the skin.
27) ***Coelogyne flaccida* Lindl.:** Paste of pseudobulbs is a remedy for treating headaches and fever, and its juice is a cure for indigestion.
28) ***Coelogyne fuscens* Lindl.:** The paste of pseudobulb is administered orally to treat stomachache.

29) ***Coelogyne gardneriana* (Lindl.) Rechb.f.:** The plant extract has shown antibacterial activity against *E. coli* and *S. typhi* and, hence, it is useful for treating diarrhea and typhoid fever.
30) ***Coelogyne nitida* (wallich. ex Don) Lindl. (=C. ochracea lindl.):** The juice of pseudobulb is a treatment for stomachache and other gastric problems.
31) ***Coelogyne ovalis* Lindl.:** It is also called "Jivanti" and is a remedy for cough, urinary infections, and eye problems. It is also useful as an aphrodisiac.
32) ***Coelogyne prolifera* Lindl.:** Paste of pseudobulb is applied to relieve headaches, fever, and also to treat burns and boils. Paste is also a cure for backache.
33) ***Ceologyne punctata* Lindl.:** Dried pseudobulbs are made into a powder which is a remedy for burn, and its spots.
34) ***Ceologyne stimulans* Rolfe (=C. *aloifolium Swartz.*):** Pseudobulbs contain mucilaginous substance and are hence used as salep.
35) ***Ceologyne stricta* (D.Don) Schltr.:** Paste of pseudobulbs is a treatment for fever and headache in NE states of India.
36) ***Conchidium muscicola*** (Lindl.) Lindl.: The whole plant is used as a treatment for cardiac, respiratory, and nervous system disorders.
37) ***Corymborkis veratrifolia* (Reinwardt.) Bl.:** Juice of fresh leaves is useful as an emetic to cure fever in children.
38) ***Cremastra appendiculata* (D.Don) Makino**: Root paste is a cure for dental cavities. Tuber is a treatment for abscesses, freckles, and scrofula and is also used as an antidote for snake poison.
39) ***Cymbidium aloifolium* (L.) Sw.:** The leaves are widely used for the treatment of boils and fever by the Naga tribes. The rhizome and roots are made into a paste with ginger to cure vomiting and loose motions. Also used to treat chronic illness, weak eyesight, vertigo, and paralysis. Plant seeds are used to heal wounds. The whole plant is also used as a tonic. 1–2 drops of leaf juice are used to treat otitis and internal inflammation.
40) ***Cymbidium devonianum* Lindl. ex Paxton:** Decoction of the whole plant is taken orally to cure cough and cold. Root paste treats boils.
41) ***Cymbidium elegans* Lindl.:** Fresh juice of the whole plant is applied on deep wounds to stop bleeding and spreading infection.
42) ***Cymbidium ensifolium* (L.) Sw.:** Root decoction is a cure for gonorrhea and decoction of flowers is a remedy for sore eyes.
43) ***Cymbidium gigantium* wall. ex Lindle.:** juice from leaves is used to stop bleeding immediately from cuts and wounds. The plant contains triterpenoid glucoside and cymboside.
44) ***Cymbidium hookerianum Rechb. f. (=C. grandiflorum Guill.):*** Seeds are used to stop bleeding from cuts, wounds, and skin injuries.
45) ***Cymbidium iridioides* D.Don:** Fresh leaves, crushed, are used to stop bleeding immediately. The whole plant in the form of powder acts as a tonic.
46) ***Cymbidium macrorhizon* Lindl.:** Decoction of rhizome is a treatment for boils and is also used as a diaphoretic (Nongdam, 2014).
47) ***Cyperipedium calceolus* (Willd.) Correl. (=Cremastra parvifolium pubescens (Willd.) Knight; C. pubescens (Willd.) Knight.:** Roots are diaphoretic, antispasmodic, hypnotic, sedative, nervine, and tonic. They are used as a treatment for diabetes, diarrhea, dysentery, paralysis, impotence, convalescence, and malnutrition. The orchid can be used as a substitute for sugandhabala (*Valeriana officinale*), though it is somewhat inferior in medicinal quality.
48) ***Cyperipedium Cordigerum* D.Don**: Roots are eaten as vegetables. They are useful as a tonic as well.
49) ***Cyperipedium elegans* Rechb. f.:** Roots are nervine tonic and are beneficial in hysteria, insanity, and epilepsy. They are a remedy for spasms and rheumatism.

50) ***Cyperipedium himalaicum* (Rolfe) Kranz.:** The orchid is a cure for urine retention, stones, etc. It is also useful in cardiac and respiratory problems, including cough.
51) ***Dendrobium amoenum* Wall. ex Lindl.:** The paste of fresh pseudotubers is applied on dislocated bone for resettling and curing skin burns.
52) ***Dendrobium crepidatum* Griff.:** Pseudobulbs paste is a remedy for treating bone fractures and dislocated bones.
53) ***Dendrobium densiflorum* Wall. (=*D. thyrsiflowrum* Reich.f.):** The whole plant is used to treat bone fractures. The leaf contains ayapin, scopoletin, and 6, 7- dimethoxy coumarin. The paste is also used for the treatment of boils, pimples, and skin eruptions. Powder of dried pseudobulbs is used as a tonic.
54) ***Dendrobium chrysanthum* Wall. ex Lindl**.: Leaf is a cure for fever and skin diseases. Stem is useful as a tonic to strengthen the stomach and immune system and to increase fluids of body system.
55) ***Dendrobium chrysotoxum* Lindl.:** Leaf extract cures fever. Stem and flower decoction is used as a tonic.
56) ***Dendrobium eriaeflorum* Griff.:** Dried flower is used as a tonic. Paste of pseudobulb treats dislocated and fractured bones.
57) ***Dendrobium denudans* D.Don:** Stem is a nutritious tonic for old people and children. It is also a treatment for cough and cold, nasal block, and tonsilitis.
58) ***Dendrobium farmeri* Paxton:** The orchid has shown a broad spectrum antibiotic activity against *klebsiella pneumoniae*, *Escherichia coli,* and *Salmonella typhi*.
59) ***Dendrobium falconeri* Hook**.: The orchid is useful in treating lung cancer and is also taken as a tonic.
60) ***Dendrobium fimbriatum* Hook.:** The whole plant is used to treat liver disorders and nervous debility. Leaf used in bone fracture. Pseudobulb is considered an aphrodisiac, stimulant, and a demulcent.
61) ***Dendrobium fugax* Rechb. f.:** The powder of dried whole plant is used as a tonic and for debility.
62) ***Dendrobium heterocarpum* Wall. ex Lindl.:** Pseudobulb is a treatment for dislocated and fractured bones.
63) ***Dendrobium hookerianum* Lindl.:** Flowers are cooked as a vegetable and also to make yellow dye in different regions of Himachal Pradesh.
64) ***Dendrobium jenkinsii* Wall. ex Lindle.:** Fresh and dried stems are used for treating gastritis, thirst in fever, and dry eye conditions. They are also useful as an aphrodisiac, analgesic, and as stomachic, as well as for prolonging age.
65) ***Dendrobium longicornu* Lindle.:** Plant juice is taken to relieve fever. Boiled roots are a treatment for livestock cough.
66) ***Dendrobium macaraei* (Lindle.) Seidenf.:** Paste of the whole plant is a treatment for snake bites. It is also used to treat asthma, bronchitis, fever with throat trouble, as an aphrodisiac, and as a demulcent.
67) ***Dendrobium moniliforme* (L.) Sw.:** Paste of pseudobulb is anti-inflammatory.
68) ***Dendrobium monticola* P.F. Hunt and Smmerh.:** Paste of pseudobulb is used for treating boils, pimples, and other skin eruptions.
69) ***Dendrobium moschatum* Lindl.:** Paste of pseudobulbs is used to treat dislocated and fractured bones. Leaf juice is a remedy for earache.
70) ***Dendrobium nobile* Lindl.:** The paste of pseudobulb is a remedy for eye infections and to soothe burn injuries in the Khasi hills. Fresh and dry stems are used in Chinese traditional medicines for longevity, as an aphrodisiac, stomachic, and analgesic. It is also useful for treating pulmonary tuberculosis, flatulence, general weakness, dyspepsia, night sweats, fever, and anorexia in Vietnam.

71) ***Dendrobium transparens* Wall. ex Lindl.:** Paste of pseudobulb is used to treat bone fractures and dislocated bones.
72) ***Dienia cylinderostachya* Lindl.:** Powder of dried pseudobulb is used as a tonic and as a remedy to improve kidney function.
73) ***Epipactus helleborine* (L.) Crantz.:** Tubers are used to treat gout, stomachache, and headache. The whole plant is used in treating nervous problems.
74) ***Eria spicata* (D.Don) Hand. Mazz. (=Pinalia spicata (D.Don)) S.C. Chen and J.J. Wood:** Stem paste is taken orally for stomachache and is also applied externally to treat headache.
75) ***Eria andamanaca* Hook. F.:** Pseudobulbs are nutritious and are consumed.
76) ***Eria pannea* Lindl. (= Mycaranthes pannea (Lindl.)) S.C. Chen and J.J. Wood:** Decoction of roots and leaves is mixed with bathing water to treat body pains.
77) ***Eulophia epidendraea* Fischer (=E. Virens R. Br.):** The orchid is mainly found in dry areas of west Bengal, Madhya Pradesh, and in states of the Deccan Plateau. The tubers are used for treating intestinal worms.
78) ***Eulophia explanta* Lindl.:** Pseudobulbs are beneficial to treat impotence, sexual weakness, spermatorrhea, and as an external application to treat cuts and wounds.
79) ***Eulophia herbacea* Lindl.:** It is known as "Vana Singhara" in Maharashtra. The pseudobulbs are mucilaginous and are used as tonic or salep.
80) ***Eulophia spectabilis* (Dennst.). Suresh:** Tubers are used as an appetizer, aphrodisiac, tonic, and a blood purifier. They are also beneficial for treating glandular tuberculosis of the neck, tumors, and bronchitis. Leaf paste is a remedy for skin disease. Tuber decoction is a treatment for pinworm and roundworm infestation in children.
81) ***Eulophia pratensis* Lindl.:** In Maharashtra, it is considered a substitute for "Shatavari" (*Asparagus racemasus*). Pseudo tubers are used to treat scrofulous glands in the neck, hard swellings, and tumors, as well as as a wormifuge.
82) ***Galeris stracheyi* (Hook.f.) P.F. Hunt:** Tubers are useful as tonic and as a remedy for treating headaches.
83) ***Gastrodia elata* Blume:** A fine powder of stems is inhaled to relieve allergies, dizziness, and epilepsy. The leaf paste is useful to treat sinusitis and migraines.
84) ***Geodorum candidum* Wall.:** It is a valuable remedy for treating scorpion stings.
85) ***Geodorum densiflorum* (Lam.) Schl.:** Tubers are used to treat a viral fever (ephemeral fever) of bovine and are useful to treat insect bites.
86) ***Goodyera repens* (L.) R. Br.:** Tuber paste is externally used to treat syphilis. Plant extract is taken as a blood purifier. Roots and leaves are used for treating female diseases and stomach and bladder diseases. Leaves are chewed and applied to snake bites. Crushed leaves are a cure for skin rashes in children.
87) ***Goodyera schlectandaliana* Rechb.f.**: Stem and leaf paste is used as a blood purifier and is beneficial to treat acne, pruritis, psoriasis, and skin fungal infections (e.g., hair dandruff). The whole plant is used as a tonic, as a cure for internal injuries, and to improve blood circulation.
88) ***Gymnadenia orchids* Lindl.:** Pseudobulbs are used as aphrodisiac and tonic in Uttarakhand. They are also used to heal cuts and wounds and as a cure for gastric ailments and liver and urinary diseases.
89) ***Habenaria commelinifolia* (Roxb.) Lindl.:** The whole plant is used as salep and is useful as a tonic, particularly in paralytic cases, diabetes, phthisis, and chronic cases of diarrhea and dysentery. The tubers are also useful for fever and nose bleeding. The whole plant is a remedy for snake bites.
90) ***Habenaria dentata* (Sw.) Schltr.:** The whole plant is used as an analgesic, aphrodisiac, disinfectant and as antirheumatic. It is also useful for treating urinary and orthopedic problems.

91) ***Habenaria furcifera* Lindl.:** Paste of tubers is used to treat cuts and wounds as well as to treat poisonous insect bites.
92) ***Habenaria crinifera* Lindl.:** It is known as "Vana Sirso" in Sikkim. The whole plant is useful to treat headaches.
93) ***Habenaria diphylla* Dalz.:** Tubers are used as a substitute for padam charini (*Nervillea aragoana* Goud.) and are useful to treat dysuria, urinary stone, colic, cough, asthma, and poisoning.
94) ***Habenaria goodyeroides* D. Don:** A terrestrial orchid generally used as a substitute for "Rddhi" and "Vrddhi".
95) ***Habenaria Longicornicudata* Grah. (=H. longicalcarata Rich.):** Tubers are useful for treating swellings and internal pain in Maharashtra, Andhra, Telangana, and Madhya Pradesh.
96) ***Habenaria marginata* colebr.:** Tubers are used as tonic and to cure flatulence. They are also a remedy for wound healing.
97) ***Habenaria pectinata* (Sm.) D.Don:** Leaf paste is applied on snake bites, and the paste of tubers mixed with "Khoya" and powder of greater cardamom is taken to relieve joint pains.
98) ***Habenaria polytricha* (Hook.f.) Pradhan:** Tubers are used as substitute for "Rddhi and Vrddhi".
99) ***Habenaria pantlingiana* Kraenzl.:** Tuber paste and its extract is useful to treat snake bite venom.
100) ***Habenaria susannae* (L.) Lindl.:** Tuber paste is useful to treat blebs (large blisters) or bullae (fluid-filled red blisters on the palms of the hand).
101) ***Hermenium monorchis* (Thunb. ex Sw.) R. Br.:** Roots are used as a tonic.
102) ***Herminium lanceum* (Thunb. ex Sw.) Vuijk.:** Decoction of the whole plant is a remedy for suppressed urination.
103) ***Limodorum spathulatum* (L.) Willd.:** The orchid is known as "Svarna pushpa". Its flowers mixed with flowers of Habenaria disphylla are useful for treating asthma.
104) ***Liparis nervosa* (Thunb.) Lindl.:** Tubers are beneficial treatment for stomachache and malignant ulcers.
105) ***Liparis odorata* (Willd.) Lindl.:** Leaf juice is a treatment for cancerous ulcers, gangrene, and skin burns. Stem is a remedy for throat cancer.
106) ***Luisia tenuifolia* Bl.:** The whole plant is an emollient and is applied in the form of a poultice for treating boils, tumors, and abscesses.
107) ***Luisia trichorhiza* (Hook.) Bl.:** The paste of tubers is used to treat muscular pain. Plant extract exhibited a high degree of inhibition of the growth of *Klebsiella pneunoniae* and *Salmonella typhi* in test experiments.
108) ***Luisia zeylanica* Lindl.:** Leaf juice is a treatment for boils, chronic wounds, and skin burns.
109) ***Luisia tristris* (G. Forst.) Hook. f.:** Leaf juice is useful for healing chronic wounds, boils, and skin burns.
110) ***Malaxis cylindrostachya* (Lindl.) O. Kuntze.:** A terrestrial orchid growing between 2,000 m and 3,000 m elevations is used as a tonic.
111) ***Malaxis rheedii* Sw.:** It is generally used as a substitute for "Malaxis muscifera" and "Crepidium acuminatum" which are said to pacify vata, pitta, raktapitta (bleeding disorders), and raktavata (gout) doshas, and is used to cure diseases caused by them and is a tonic and also as an aphrodisiac.
112) ***Malaxis sapropayta* (Kings Prantl.) Tong and Wang:** In NE region of India, it is locally called "walkhum", and its bulbs are used to heal burn injuries.
113) ***Monomeria barbata* Lindl.:** Tubers are useful for treating cough, pulmonary tuberculosis, and trauma (great shock after an accident).
114) ***Neottianthe ccucullata* (L.) Schltr.:** Rhizome is useful as a tonic.

115) ***Nervilia aragoana* Goud. (Padamcharini):** The tuber is hot, bitter, and astringent. It is a treatment for morbid vata and kapha doshas and is useful to treat urinary stones, dysuria, colic, cough, asthma, and as an antidote to poisoning conditions.
116) ***Nervilia macroglossa* (Hook.f.) schltr.:** Tuber is used to treat male sterility.
117) ***Oberronia caulescens* (Lindl.):** Tuber is a useful remedy for liver disorders.
118) ***Oberronia falconeri* (Hook.f.):** Whole plant paste is a treatment for bone fractures.
119) ***Oberronia pachyrachis* Rchb.f. exHook.f.:** The extract of the whole plant has shown a pronounced effect against *Klebsiella pneumoniae* and *S. aureus*.
120) ***Orchis laxiflora* Lam. (=*O. ensifolia* Vill.):** It is popularly called the "Marsh orchid or oriental salep" and is treatment for diarrhea, bronchitis, and convalescence. The tubers are a rich source of mucilage (salep).
121) ***Otochilus albus* Lindl.:** The whole plant powder is used as a tonic.
122) ***Otochilus lancifolius* Griff.:** Pseudobulb is used to treat dislocated and fractured bones.
123) ***Otochilus porrectus* Lindl.:** The whole plant powder is used as a tonic and also to treat sinusitis and rheumatism.
124) ***Paphiopedalium hirsutissimum* (Lindl. ex Hook.) Stein.:** Leaves are used to treat scabies, and roots are a remedy for ringworms in Nagaland.
125) ***Papilionanthe teres* (Roxb.) schltr.:** Whole plant paste is a treatment for dislocated bones.
126) ***Peristylis constrictus* (Lindl.) Lindl.:** Fresh root paste is applied on boils as a treatment.
127) ***Peristylis stocksii* (HK.f.) Kranz.:** It is known as "Salam Kanda". The root is used to treat debility and is considered a tonic and nutritious by the people of Madhya Pradesh.
128) ***Phaius tankervilliae (Banks)Blume:*** Tuber is used as a tonic. Paste of tuber is used to treat swellings of gout, to reduce pain of fractured bone, rheumatic or gouty swellings of hands and legs, and redness of swellings. It is taken orally to treat dysentery. A poultice of tubers and leaves is a remedy for infected wounds and boils as well as to relieve the pain of abscesses.
129) ***Pholidota chinensis* Lindl.:** Pseudobulb is a remedy for toothache, stomachache, and inflammatory conditions.
130) ***Pholidota articulata* Lindl.:** The whole plant is useful as a tonic. Powder of dried roots is used to treat cancer. Juice of fruits is a treatment for skin ulcers and skin eruptions. It is also a substitute for "Jivanti" (*Flickingeria fimbriata*).
131) ***Pholidota imbricata* Lindl.:** Paste of pseudobulb is a treatment for fever, pain, and swelling of arthritis. Powder is useful as a tonic. Juice of pseudobulb is used to relieve painful conditions of rheumatism.
132) ***Pholidota pallida* Lindl.:** Pseudobulb is used to treat sleep disorders and root juice treats nasal, abdominal, and rheumatic pains in Sikkim. It is used to stop bleeding from cuts.
133) ***Pinalia spicata* (D.Don) S.C. Chen and J.J. Wood:** The stem paste is taken internally for treating stomachache and also applied externally to treat headache.
134) ***Plantathera clavigera* (Lindl.) Dandy (= Herminium clavigerum, new name):** Tubers are known as "Bhyathui" in Sikkim, and these are eaten as a source of nutrition.
135) ***Plantathera latilabris* Lindl.:** Pseudobulbs are used as blood purifiers.
136) ***Plantathera Sikkimensis* (Hook.f.) Kraenzin:** Pseudobulbs are used to relieve the pain of the naval point, abdomen, and rheumatism.
137) ***Ponerorchis chusua* Bl.:** Tubers are used to treat diarrhea, dysentery, and chronic fevers.
138) ***Pleione humilis* (Sm.) D. Don:** Dried pseudobulb powder is a tonic, and paste is used to heal cuts and wounds.
139) ***Pleione maculata* (Lindl.) Lindl.:** The rhizome is used to treat liver and stomach diseases.
140) ***Pleione praecox* (Sm.) D.Don:** Pseudobulb treats cuts and wounds, and powder is used as a tonic.
141) ***Renanthera imschootiana* Rolfe:** Leaf paste treats skin diseases.

142) ***Rhynchektylus retusa* (L.) Bl**.: It is a substitute for "Rasna". Leaves are used to treat rheumatic pain. Roots juice is applied externally for healing cuts and wounds. Khasis uses the whole plant to treat skin diseases. The plant extract has shown activity against *E. Coli* and *Bacillus subtilis*.
143) ***Satyrium nepalense* D. Don:** It is also called "Salab miskri". Tubers are used internally to treat malaria and dysentery. They are also used as a tonic.
144) ***Samitinandia micrantha* (Lindl.) Holttum.:** Root powder is used as a tonic, and stem has shown antibacterial properties.
145) ***Spiranthes sinensis* (Pers.) Ames:** Tubers are used as a tonic. Decoction of the plant treats intermittent fever. Paste of stem and roots is a treatment for skin sores.
146) ***Thunia alba* (Lindl.) Richb.f.:** Paste of the whole plant is a treatment for dislocated bones.
147) ***Tropidia curculigoides* Lindl.:** Decoction of roots is used to treat diarrhea and of the whole plant along with *Ardisia Colorata* Roxb. wich is called "Bol-Simbal" in Garohills and "Rulthlu" in Mizoram, is a cure for malaria in the cold stage.
148) ***Trudelia cristata* (Lindl.) Senghas ex Roeth. (=Vanda cristata Lindl):** It is also known as "Hadajoda". Root paste is a treatment for bone dislocation and bone fracture.
149) ***Vanda Coerulea* Griff. (ex.) Lindl.:** Leaves are useful for treating diarrhea and dysentery and is also used as an expectorant.
150) ***Vanda roxburghii* R. Br. (= *Vanda tessellata*):** Various parts of the plant are used traditionally as antipyretic and are useful in treating bronchitis, dyspepsia, inflammations, piles, and diseases of the nervous system. It is also useful as a laxative and to treat hiccough, piles, boils, and scorpion stings.
151) ***Vanda testacea* (Lindl.) Rechb.f**.: Leaf juice treats earache. Leaves are antiviral and anticancer.
152) ***Zeuxime longilabris* (Lindl.) Trimen:** The whole plant is a treatment for whooping cough.
153) ***Zeuxine strateumatica* (L.) Schltr:** Roots and tubers are dried and made into powder for use as a tonic.

4.4 CONCLUSION

Orchids are an integral part of the daily lives of forest dwellers and different tribes around the globe. These plants are used extensively because of their ornamental, religious, and medicinal values. It has been observed that the medicinal values of orchids have been well known to the local population since the early Harappan times and the early Vedic period. Various texts from the period have mentioned the uses of these plants, indicating their known significance as a potent medicine. Different species of orchids were used by indigenous populations and ayurvedic practitioners as a treatment or cure for different ailments such as tuberculosis, skin disorders, ulcers, digestive disorders, etc. Orchids are known to have different active chemicals that exhibit varying medicinal properties. The knowledge of orchids, if well-researched and analyzed, could be utilized in modern-day medicine.

REFERENCES

Aswandi, A., & Kholibrina, C. R. (2021, November). Ethnomedicinal properties of orchidaceae by local communities in Lake Toba region, North Sumatra, Indonesia. In *IOP Conference Series: Earth and Environmental Science* (Vol. 914, No. 1, p. 012056). IOP Publishing.

Bulpitt, C. J., Li, Y., Bulpitt, P. F., & Wang, J. (2007). The use of orchids in Chinese medicine. *Journal of the Royal Society of Medicine*, *100*(12), 558–563.

Gutiérrez, R. M. P. (2010). Orchids: A review of uses in traditional medicine, its phytochemistry and pharmacology. *Journal of Medicinal Plants Research*, *4*(8), 592–638.

Jalal, J. S., Kumar, P., Tewari, L., & Pangtey, Y. P. S. (2010). Orchids: Uses in traditional medicine in India. In *National Seminar on Medicinal Plants of Himalaya: Potential and Prospect.* Regional Research Institute of Himalayan Flora. www.researchgate.net

Nongdam, P. (2014). Ethno-Medicinal Uses of Some Orchids of Nagaland Northeast, India. *Research Journal of plant, 8*(3), 126–139.

Panda, A. K., & Mandal, D. (2013). The folklore medicinal orchids of Sikkim. *Ancient Science of Life*, *33*(2), 92–96.

Pant, B. (2013). Medicinal orchids and their uses: Tissue culture a potential alternative for conservation. *African Journal of Plant Science*, *7*(10), 448–467.

Sharifi-Rad, J., Quispe, C., Bouyahya, A., El Menyiy, N., El Omari, N., Shahinozzaman, M., Ovey, M. A. H., Koirala, N., Panthi, M., Ertani, A., Nicola, S., Lapava, N., Herrera, J., A Salazar, L., Changan, S. S., Kumar, M., & Calina, D. (2022). Ethnobotany, phytochemistry, biological activities, and health-promoting effects of the genus *Bulbophyllum*. *Evidence-based Complementary and Alternative Medicine*, *2022*.

Sharma, B. D. (2021). Human wellness through the beautiful orchids. In Indu Rani Sharma (ed) *Medicinal Plants: Herbal Wealth of India* (pp. 31–41). Agrobios Research an Imprint of *Agrobios* (India).

Sharma, B. D., & Balkrishan, A. (2005). *Vitality Strengthening Astavarga Plants (Jeevaniya & Vayasthapan Paudhe)*. Divya Yog Mandir.

Sharma, P. V. (2005). *Carak Samhita Vol. 3 (1981) and Vol. 4 (005) and Vol. 2 (2021)*. Chaukhamba Orienetalia Varansi.

Singh, A., & Duggal, S. (2009). Medicinal orchids-an overview. *Ethnobotanical Leaflets*, *2009*(3), 3.

Singh, B. (2022). Therapeutic Himalayan herbs: Folklore uses, bioactive phytochemicals, and biological activities of medicinal orchids used by nomads. *Indian Journal of Natural Products and Resources*, *13*(1), 94–104.

Singh, S., Singh, A. K., Kumar, S., Kumar, M., Pandey, P. K., & Singh, M. C. K. (2012). Medicinal properties and uses of orchids: A concise review. *Elixir Applied Botany*, *52*(2012), 11627–11634.

Tsering, J., Tam, N., Tag, H., Gogoi, B. J., & Apang, O. (2017). Medicinal orchids of Arunachal Pradesh: A review. *Bulletin of Arunachal Forest Research*, *32*(1&2), 1–16.

Varma, A. R., & Ingalhalli, R. (2022). A review article on: An overview of medicinal importance of orchids. *Acta Scientific Medical Sciences*, *6*(7), 28–36.

Yonozone, R., Bhujel, R. H., & Rai, S. (2012). Genetic resources, current ecological status and altitudes wise distribution of medicinal plants diversity of Darjeeling Himalaya of West Bengal, India. *Asian Pacific Journal of Tropical Biomedicine*, *2012*(5), 78–88.

Yonzone, R., Lama, D., Bhujel, R. B., & Samule, R. (2013). Present availability status, diversity resources and distribution of medicinal orchid species in Darjeeling Himalaya of West Bengal, India. *International Journal of Pharmacy and Natural Medicine*, *1*(1), 14–35.

5 Medicinal Values and Other Potential Uses of *Bulbophyllum* Orchids

Kiran Bisht, Aarushi Gautam, Manu Pant, Ashish Gaur, Nital Nilesh Chankeshwara, and Abhijeet Rathod

5.1 INTRODUCTION

Orchids are the highly evolved and largest groups of angiosperms, containing nearly 736 well-studied genera and 2,800 species spread throughout the globe (Chase et al., 2015). *Bulbophyllum* is one of the most represented genera, with more than 2,000 species distributed in parts of Asia and Africa. The Asian regions of China, India, Nepal, Thailand, Laos, and Vietnam are known to house abundant *Bulbophyllum* species (Lin et al., 2020). Nearly 100 species of *Bulbophyllum* have been reported alone in India, indicating its morphological and anatomical adaptations. *Bulbophyllum* sp. are epiphytic, and the habitat ranges from subtropical to alpine forest (Hosseini and Dadkhah, 2016). Apart from the ornamental value of the species, it has been extensively used in herbal medicinal preparations and traditional medicines. The active phytochemicals obtained from the plant contain various healing properties and have been utilized by various tribes. This has encouraged research investigations to explore and confirm the phytochemistry of the genus and use it for modern medicine (Hossain, 2011).

5.1.1 Morphology and Distribution

The *Bulbophyllum* genus is mainly epilithic with ariel roots, most commonly fibrous or filamentous. The main stem is usually differentiated into pseudobulbs, and leaves are generally ovate or orbiculate. The thick, leathery leaves are arranged in fan-like patterns for efficient transpiration and light capture (Nunes et al., 2014). The inflorescence in the genera is usually racemose, with fused sepals and petals forming a perianth, serving as a protective sheath (Teoh, 2016). Plants generally require warm temperatures for growth and a minimum temperature of around 22^{0}C, requiring moderate to bright sunlight conditions.

Bulbophyllum has global distribution centered in Papua New Guinea, but spans Southeast Asia, Madagascar, Africa, Australia, India, and tropical South and Central America (Gamisch and Comes, 2019). With over 2,000 species of global presence, 100 species thrive in India alone. China hosts more than 105 species, Madagascar has over 200, and more than 122 are from Vietnam (Abraham, 1981). The global distribution of the Bulbophyllum orchid is shown in Figure 5.1. A few common *Bulbophyllum* species are *Bulbophyllum lobbii*, *Bulbophyllum echinolabium*, *Bulbophyllum falcatum*, *Bulbophyllum longissimum*, *Bulbophyllum fascinator*, *Bulbophyllum flabellum-veneris*, *Bulbophyllum lasiochilum*, *Bulbophyllum dearie*, *Bulbophyllum makoyanum*, and *Bulbophyllum medusae*.

DOI: 10.1201/b23375-5

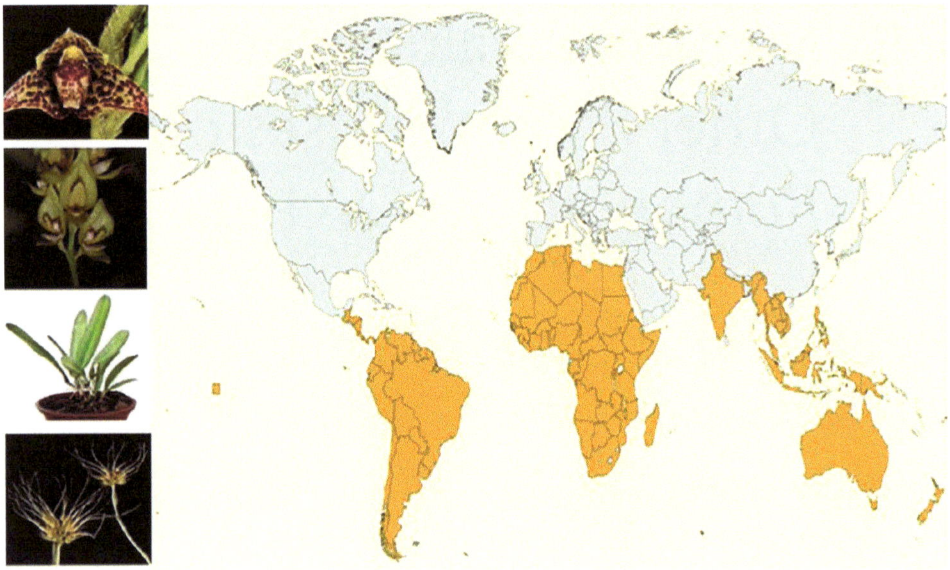

FIGURE 5.1 The geographical distribution of *Bulbophyllum* species.

5.2 PHYTOCHEMISTRY

Various research suggests that the presence of active components in *Bulbophyllum* is responsible for the pharmacological activity. The great diversity of the phytochemicals in the genus make it desirable for use in various medicinal preparations and to be explored for further research. It is suggested that the quality and quantity of the chemical composition of the *Bulbophyllum* species is unique and varies from species to species (Holttum, 1964). However, various external factors such as temperature, geography, soil, and climate can also result in slight variations in the active components within the species (Amirkia and Heinrich, 2014).

Bulbophyllum albidum, a plant species endemic to the Western Ghats of India, has been investigated by Lalitharani et al. (2011) for its phytochemical composition (Dash et al., 2008). The analysis revealed the presence of diverse compounds, including alkaloids, tannins, catechins, phenols, xanthoproteins, glycosides, flavonoids, and saponins. *B. kaitense*, a plant endemic to Southern India, was investigated in a study by Kalaiarasan et al. (2012a) with the use of Gas Chromatography-Mass Spectrometry (GC-MS) analysis. The ethanolic extract included 13 bioactive substances, including butyl 2-methyl propyl ester (3.45%), dodecyl phthalate (3.52%), methyl 3-bromo-1-adamantaneacetate (13.03%), squalene (23.22%), and others. *Bulbophyllum kwangtungense*, a native of China and Japan, reported the presence of various phytochemicals, including 7,8-dihydro-5-hydroxy-12,13-Methylenedioxy-11 methoxyldibenz[b,f] oxepine, cumulatin, 7,8-dihydro-4-hydroxy-12,13-methylenedioxy-11-Methoxyldibenz[b,f]oxepine,7,8-dihydro-3-hydroxy-12,13-methylenedioxy 11methoxyldibenz[b,f]oxepine, plicatol B, and densiflorol A. *Bulbophyllum sterile* is another native Indian plant that contains a wide range of phytochemicals, including tannins, glycosides, alkaloids, coumarins, phenols, and saponins. *Bulbophyllum leopardinum* is native to Bhutan, Nepal, India, northern Vietnam, and China. Numerous phytochemicals, such as phenanthrene and bulbophyllanthrene, have received scientific interest (Majumder et al., 1985). *Bulbophyllum reptans* is found in India, Bhutan, Nepal, Southeast China, Hainan, Guangxi, Yunnan, Vietnam, Myanmar, Sikkim, etc. It also contains various phytochemical constituents, such as cologne, flavanthrin, cirropetalanthrin, confusarin, gynopuscinblestriaren A, reptanthrin, and isoreptanthrin (Majumder et al., 1999). Plant species *Bulbophyllum retusiusculum*, which

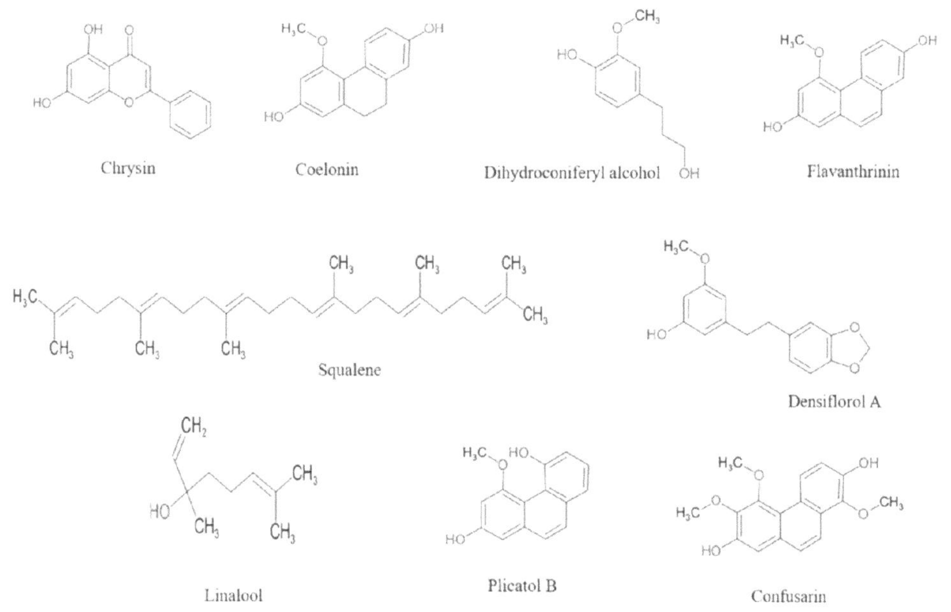

FIGURE 5.2 Phytochemicals present in different *Bulbophyllum* species.

grows throughout Southeast Asia, including Hainan, Guangxi, Yunnan, India, Myanmar, Bhutan, Sikkim, Vietnam, Nepal, and China (Teoh, 2016), includes retusiusine A, retusiusine C, retusiusine B, retusiusine E, retusiusine F, retusiusine G, dihydroconiferyl dihydro-pcoumarate, methyl 3-(4-hydroxyphenyl) propionate, 3-(4-hydroxyphenyl)-propanoic acid, dihydroferulic acid, methyl 3-(4-methoxyphenyl) propionate, 3-(3,4-methoxyphenyl)-2-propenal, trans-p-coumaric acid, and dihydroconiferyl alcohol. *Bulbophyllum neilgherrense,* native to Karnataka, India (Harshitha et al., 2013), contains a variety of phytochemicals such as phenols, saponins, tannins, alkaloids, and glycosides; some of these are shown in Figure 5.2.

5.3 ETHNOPHARMACOLOGY

Many traditional medicines have orchids as an important ingredient. With the advancement of science and technology, there are several research studies in modern medicine suggesting the potency of active constituents of orchids in treating various health issues. Among all the categories of medicinal orchids, *Bulbophyllum* is one of the most explored ones. *Bulbophyllum* species have been used to treat various life-threatening conditions and, hence, the most explored genus. The phytochemicals present in *Bulbophyllum* are known to interact with the biomolecules in the body and produce the desired effects. The species are known to exhibit antimicrobial, anticancer, and anti-inflammatory effects (Figure 5.3, Table 5.1). In China, *Bulbophyllum* orchids have been employed in traditional medicine for over 2,000 years. Records from ancient Chinese texts mention the use of orchids as a treatment for tuberculosis and severe respiratory disorders *B. kwangtungense*, *B. anderson, B. odoratissimum*, and *Pholidota* spp. were utilized by the Tujja and Miao tribes to treat fractures, tuberculosis, impotence, and pain of hernias (Qu et al., 2006). Women in Myanmar are known to grind pseudobulbs from different *Bulbophyllum* species to make hair tonics and shampoos. This concoction is known to treat dandruff, ensure hair shaft development, and enhance the color and texture of hair. These orchids have been an important part of Ayurvedic medicine in India as well, where they have been used to treat diseases such as diabetes, tuberculosis, colds, and coughs (Kurzweil and Lwin., 2014).

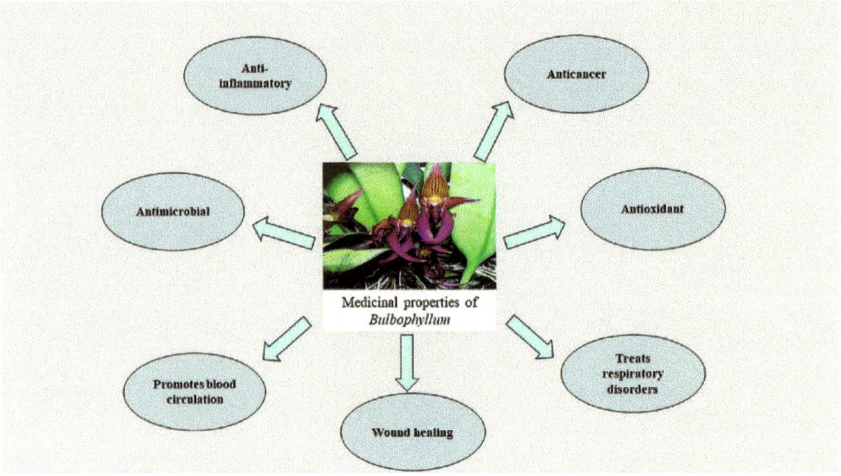

FIGURE 5.3 General medicinal properties of *Bulbophyllum* species.

TABLE 5.1
Phytochemicals and Medicinal Properties of *Bulbophyllum* Species

S. No.	Scientific Name	Location	Phytochemicals	Medicinal Properties	References
1	*Bulbophyllum albidum*	Western Ghats of India	Alkaloid Tannin Catechin Phenol Xanthoprotein Glycoside	Strengthening of a weak uterus for conception.	Lalitharani et al. (2011)
2	*Bulbophyllum kaitense Rechib*	Southern India	Squalene 3-Buten-2-ol Methyl 3-bromo-1-adamantane acetate 3,4-Hexanediol, 2,5-dimethyl- 2,4-Dimethylcyclopentanol Didodecyl phthalate	Stabilization of lysosomal membranes, preventing the release of inflammatory compounds.	Kalaiarasan et al. (2012)
3	*Bulbophyllum kwangtungense*	China, Japan	7,8-dihydro-5-hydroxy-12,13-methylenedioxy-11-methoxyldibenz[b,f]oxepin 7,8-dihydro-4-hydroxy-12,13-methylenedioxy-11-methoxyldibenz[b,f]oxepin 7,8-dihydro-3-hydroxy-12,13-methylenedioxy-11-methoxyldibenz[b,f]oxepin Cumulatin Densiflorol A Plicatol B	Treat pulmonary tuberculosis, promote the production of body liquid and reduce fever.	Wu et al. (2006) and Gutiérrez (2010)

(Continued)

TABLE 5.1 *(Continued)*
Phytochemicals and Medicinal Properties of *Bulbophyllum* Species

S. No.	Scientific Name	Location	Phytochemicals	Medicinal Properties	References
4	*Bulbophyllum odoratissimum*	China, Nepal, Sikkim, Bhutan, India, Burma, Thailand, Laos, and Vietnam	3-[2-(7-methoxy-benzo[d][1,3] dioxol-5-yl)ethyl] phenol 6-(3-hydroxy-phenethyl) benzo[d][1,3]dioxol-4-ol Chrysin Pinobanksin Bulbophythrins A Bulbophythrins B	Treat tuberculosis, chronic inflammation, and fracture.	Pant and Raskoti (2013), Majumder and Sen (1991), Xu et al. (2009), and Teoh (2016)
5	*Bulbophyllum sterile*	India	Glycosides Alkaloids Coumarins Tannins Saponin	Anticancer properties.	Biswas et al. (2016), Joseph et al. (2018)
6	*Bulbophyllum neilgherrense*	Karnataka, India	Saponin Phenols Tannins Alkaloids Glycosides	Analgesic activity, moderate anti-inflammatory activity.	Harshitha et al. (2013), Nair et al. (2018)
7	*Bulbophyllum retusiusculum*	Taiwan, Hainan, Sichuan, Yunnan, Myanmar, Thailand, Indochina, Peninsular Malaysia, Nepal, Sikkim, and Bhutan	Retusiusine A, B, C, E, F, G Trans-p-coumaric acid Dihydroconiferyl alcohol Epemeranthol-A	Antibacterial, antifungal properties.	Fang et al. (2018, 2021), Teoh (2016)
8	*Bulbophyllum pectenveneris*	Fujian, Taiwan, Hong Kong, Guangxi, Hainan, Laos, and Vietnam	Alkaloids Flavonoids Terpenoids Phenols	Reducing blood stasis, promoting circulation, anti-inflammatory.	Teoh (2016)
9	*Bulbophyllum siamense*	Thailand	Alkaloids Flavonoids Terpenoids Phenolic	Leaves are used to treat burns.	Teoh (2016), Chuakul (2002)
10	*Bulbophyllum cariniflorum*	Northern Thailand, Northeast India, Bhutan, and Nepal	Alkaloids Flavonoids Terpenoids Phenols	Induce first-trimester abortion.	Teoh (2016), Dash et al. (2008)
11	*Bulbophyllum leopardinum*	Nepal, Bhutan, India, China, and northern Vietnam	Phenenthrine Bulbophyllanthrin	Treat burns on skin.	Majumder et al. (1985), Pant and Raskoti (2013), Teoh (2016)

(Continued)

TABLE 5.1 *(Continued)*
Phytochemicals and Medicinal Properties of *Bulbophyllum* Species

S. No.	Scientific Name	Location	Phytochemicals	Medicinal Properties	References
12	*Bulbophyllum reptans*	Hainan, Guangxi, Yunnan, southeast Xizang, Vietnam, Myanmar, Sikkim, Bhutan, Nepal, and India	Coelonin Flavanthrin Confusarin Gynopusinblestriarene A Cirrhopetalanthrin Reptanthrin Isoreptanthrin	Cures dry throat, reduces hemoptysis, increases appetite, and enhances healing and fractures.	Majumder et al. (1999), Teoh (2016)

5.3.1 Medicinal Properties

5.3.1.1 Antimicrobial Properties

Many *Bulbophyllum* species possess antimicrobial and antifungal properties. A study conducted by Priya and Krishnaveni (2005) reported the antibacterial properties of *B. neilgherrense* against five strains of bacteria (*Bacillus pumilus*, *Escherichia coli*, *Pseudomonas aeruginosa*, *Pseudomonas putida,* and *Staphylococcus aureus*). Pseudobulbs and leaves were selected for the aqueous, chloroform, and ethanolic extracts. Bacterial strains *E. coli*, *P. aeruginosa,* and *S. aureus* were found susceptible to the ethanolic extract of the leaves, while the ethanolic extract of pseudobulb was more effective against *P. aeruginosa* and *P. putida*.

The solvent extract (chloroform, petroleum ether, water, and ethanol) from *B. kaitense* showed antimicrobial and antifungal activities. Enhanced antifungal properties were seen in chloroform and ethanol extracts, while improved antibacterial properties were reported in aqueous, chloroform, and ethanol extracts (Kalaiarasan et al., 2012a). Kalaiarasan et al. (2012) also observed *B. kaitense* stem extract's antifungal activity against ten bacterial and fungal strains. Marasini and Joshi (2012) reported moderate antibacterial activity of *B. affine* against *Staphylococcus aureus*. Newly discovered phenylpropanoids in tubers of *B. retusiusculum* have been tested for antimicrobial activity against *B. subtilis*, *E. coli,* and *Candida albicans* with nystatin (4μg/ml) and kanamycin (4μg/ml) as positive controls for estimating the antifungal and antibacterial capabilities respectively (Fang et al., 2018). The study reported antifungal activity of retusiusine B was observed against *C. albicans* (16μg/ml), and average antibacterial properties of (±)- retusiusine C enantiomers were observed against *B. subtilis* (64 μg/ml).

5.3.1.2 Anti-Inflammatory Properties

Bulbophyllum orchids are known to possess anti-inflammatory and analgesic properties. The extract of *B. kaitense* pseudobulb (aqueous, ethanolic, petroleum ether, and chloroform) exhibited anti-inflammatory activity utilizing the human red blood cell (HRBC) method. Ethanolic extract of the pseudobulbs was found to be the most effective, while aqueous, chloroform, and petroleum ether extracts showed several anti-inflammatory properties (Kalaiarasan et al., 2012). Flavonoids such as quercitin and chrysin, tannins, phenolics, and calcium are reported to be found in pseudobulbs. Kumari et al. (2013) reported that chrysin (flavonoid), found in the pseudobulb of *B. neilgherrense* exhibited anti-inflammatory properties and inhibited the COX-2 expression and IL-6 signaling. The anti-inflammatory activity of aqueous, dichloromethane, and ethanol extract of pseudobulb, leaves, and roots of *B. scaberulum* showed particular and vital effects on COX-2 suppression. Better performance at the suppression of COX-2 was shown by the dichloromethane root extracts when compared to galantamine (a commercial drug). Meanwhile, aqueous and ethanolic extracts from leaves and pseudobulbs exhibited average COX-2 inhibition (Chinsamy et al., 2014). Pseudobulb powder of *Bulbophyllum neilgherrense* was tested for its anti-inflammatory and analgesic properties against Charles Foster rats. The results revealed that the powder mixed with honey and water was effective as an analgesic against radiant heat-induced pain, fair anti-inflammatory activity against carrageenan-induced acute inflammation, and mild activity against formalin-induced subacute inflammation and pain in rats. The powder was interpreted to be more potent against severe pain (Nair et al., 2018). Rahman and Huda (2021) conducted an experiment to identify the anti-inflammatory activity of 10 orchid species. It was reported that the highest anti-inflammatory activity was shown by *Cymbidium aloifolium* (52.6–83.3%), and adequate anti-inflammatory activity of leaf extract of *B. lilacinum* (38.1 to 66.8%) was exhibited during the egg albumin denaturation process for the inhibition of acetylsalicylic acid.

5.3.1.3 Antioxidant Properties

Due to the presence of abundant secondary metabolites in the *Bulbophyllum* species, they are believed to possess antioxidant properties that aid in healing various health issues. Reports have been published to prove that *Bulbophyllum* can be a natural source of antioxidants. An experiment conducted by Chinsamy et al. (2014) on the extracts of pseudobulbs, leaves, and roots derived from *B. scaberulum* showed average antioxidant activity (ANT) up to 100% when compared to the activity of BHT drug (95.8%) as well as was successful in inhibiting the activity of acetylcholinesterase. Phenol extracts of leaves and pseudobulb derived from *B. morphologlorum* Kraenzl and *Dendrobium* Sonia Earsakul were tested for various protein expressions. It was observed that stress-related proteins such as heat shock protein and Cu/Zn SOD were highly expressed in *B. morphologlorum*, the high expression of the antioxidant enzyme was validated by employing superoxide dismutase activity staining on native PAGE followed by LC/MS/MS (Huehne et al., 2020). Prasad et al. (2021) experimented with extracting *B. odoratissimum* leaves, pseudobulbs, and roots from the mother plant and *in vitro* raised plantlets. Upon phytochemical screening, the results revealed that every part of *B. odoratissimum* has acute antioxidant activity. The secondary metabolite production is similar in both the mother plant and the *in vitro*-raised plantlets. The antioxidant activity of *B. lilacinum,* along with nine other orchids, was tested in DPPH free radical scavenging with ascorbic acid as a standard. The results showed that all the orchids possess fair antioxidant activity (Rahman and Huda, 2021).

5.3.1.4 Anticancer Properties

Reports have been published citing the *in vitro* cytotoxicity of *Bulbophyllum* orchids. Phenanthrenes isolated from *B. incospicuum* and *B. odoratissimum* exhibited cytotoxicity in the case of cancer cell lines like human leukemia cell lines HL-60, SMMC-7721, and K562 (Tan et al., 2006; Xu et al., 2009). In a study conducted by Kalaiarasan and John (2011) on *B. kaitense* Rechin extract, various compounds were identified as possessing antimicrobial, anti-inflammatory, analgesic, antioxidant, and anticancer properties via the GC-MS method. The compounds possessing anticancer properties were squalene, tetramethoxystilbene, benzopyran, stigmasterol, phytol, and à-Bisabolol. In a study, the extract of *B. sterile* pseudobulbs and roots was successful in triggering apoptosis in human colon cancer cell lines (HCT-116) by successfully arresting the G2/M phase of the cell cycle (Biswas et al., 2016). Lusianthridin extracted from *B. blepharistes* exhibited cytotoxicity against human leukemia cell lines (HL-60) at an IC_{50} level of 26.9μM and multidrug-resistant lung cancer cell line (H69AR) at an IC_{50} level of 142.6 μM (Bhinija et al., 2021).

5.3.1.5 Other Uses

Dried root of *Bulbophyllum cariniflorum* mixed with cow milk and black pepper has been famously used for instigating abortion within two to three months of pregnancy (Leander and Lüning, 1967). Hair tonic and shampoo prepared by Myanmar women of *Bulbophyllum* species by mixing pseudobulb powder with pulverized fruit, bark, and seeds for better hair growth, to remove dandruff, and to improve hair health (Kurzweil and Lwin, 2014). The juice of *B. lilacinum* extracted from pressing the pseudobulbs when mixed with water is consumed as a cool drink. It is believed to eliminate restlessness and fatigue and keep the body fresh (Hossain, 2009).

5.4 CONCLUSION

Bulbophyllum species have been a part of traditional medicine in various parts of the world. Different species of the *Bulbophyllum* distributed throughout the wet montane to subtropical dry forests are rich in different phytochemicals. With recent advancements in techniques and research, these phytochemicals have been explored for their therapeutic effects. These chemical structures are known for their antimicrobial, anti-inflammatory, anti-cancerous, and antioxidant activity. These activities of the plant make it a desirable candidate in the pharmaceutical industry. However, there is a need to investigate the interaction of these active constituents with various biomolecules in the body. Extensive pharmacokinetic studies, including clinical studies, are needed to determine the safety and efficacy of *Bulbophyllum*-derived compounds. Important metabolic pathways and regulatory genes can be identified using proteomics and genomics, while tissue culture techniques can be used for germplasm conservation and propagation. Besides, there is also a need for effective collaborations with indigenous groups for cultivation and propagation through plant tissue culture techniques.

REFERENCES

Abraham, A. (1981). *Introduction to orchids with illustrations and descriptions of 150 South Indian orchids.* Tropical Botanic Garden and Research Institute.

Amirkia, V., & Heinrich, M. (2014). Alkaloids as drug lead–A predictive structural and biodiversity-based analysis. *Phytochemistry Letters, 10*, 48–53.

Bhinija, K., Huehne, P. S., Prawat, H., Ruchirawat, S., Saimanee, B., Mongkolsuk, S., & Satayavivad, J. (2021). The rhizome of the *Bulbophylllum* orchid is a rich source of cytotoxic bioactive compounds, potential anticancer agents. *South African Journal of Botany, 141*, 367–372.

Biswas, S., Pardeshi, R., Reddy, N. D., Shoja, M. H., Nayak, P. G., Setty, M. M., & Pai, K. S. R. (2016). *Bulbophyllum* sterile petroleum ether fraction induces apoptosis in vitro and ameliorates tumor progression in vivo. *Biomedicine & Pharmacotherapy, 84*, 1419–1427.

Chase, M. W., Cameron, K. M., Freudenstein, J. V., Pridgeon, A. M., Salazar, G., Van den Berg, C., & Schuiteman, A. (2015). An updated classification of Orchidaceae. *Botanical Journal of the Linnean Society, 177*(2), 151–174.

Chinsamy, M., Finnie, J. F., & Van Staden, J. (2014). Anti-inflammatory, antioxidant, anti-cholinesterase activity and mutagenicity of South African medicinal orchids. *South African Journal of Botany, 91*, 88–98.

Chuakul, W. (2002). Ethnomedical uses of Thai orchidaceous plants. *Mohidol University Journal of Pharmaceutical Sciences, 29*(3–4), 41–45.

Dash, P. K., Sahoo, S., & Bal, S. (2008). Ethnobotanical studies on orchids of Niyamgiri hill ranges, Orissa, India. *Ethnobotanical Leaflets, 2008*(1), 9.

Fang, Y. S., Yang, M. H., Cai, L., Wang, J. P., Yin, T. P., Yu, J., & Ding, Z. T. (2018). New phenylpropanoids from *Bulbophyllum retusiusculum. Archives of Pharmacal Research, 41*, 1074–1081.

Fang, Y. S., Yang, M. H., Wang, J. P., Fu, H., Cai, L., & Ding, Z. T. (2021). Four new phenanthrene derivatives from *Bulbophyllum retusiusculum. Fitoterapia, 152*, 104910.

Gamisch, A., & Comes, H. P. (2019). Clade-age-dependent diversification under high species turnover shapes species richness disparities among tropical rainforest lineages of *Bulbophyllum* (Orchidaceae). *BMC Evolutionary Biology, 19*, 1–16.

Gutiérrez, R. M. P. (2010). Orchids: A review of uses in traditional medicine, its phytochemistry and pharmacology. *Journal of Medicinal Plants Research, 4*(8), 592–638.

Harshitha, K., Nishteswar, K., & Harisha, C. R. (2013). Pharmacognostical and preliminary phytochemical investigations on different parts of *Bulbophyllum neilgherrense* Wight. -An orchid used in folk medicine. *Global Journal of Research on Medicinal Plants & Indigenous Medicine, 2*(4), 259.

Holttum, R. E. (1964). Orchids of Malaya. *A Revised Flora of Malaya, 1*, 759.

Hossain, M. M. (2009). Traditional therapeutic uses of some indigenous orchids of Bangladesh. *Medicinal and Aromatic Plant Science and Biotechnology*, *42*(1), 101–106.

Hossain, M. M. (2011). Therapeutic orchids: Traditional uses and recent advances—An overview. *Fitoterapia*, *82*(2), 102–140.

Hosseini, S. H., & Dadkhah, K. (2016). Intergeneric classification of genus *Bulbophyllum* from peninsular Malaysia based on combined morphological and rbcL sequence data. *Pakistan Journal of Botany*, *48*(4), 1619–1627.

Huehne, P. S., Bhinija, K., Srisomsap, C., Chokchaichamnankit, D., Weeraphan, C., Svasti, J., & Mongkolsuk, S. (2020). Detection of superoxide dismutase (Cu–Zn) isoenzymes in leaves and pseudobulbs of *Bulbophyllum morphologlorum* Kraenzl orchid by comparative proteomic analysis. *Biochemistry and Biophysics Reports*, *22*, 100762

Joseph, M., Jose, L., & Sequeira, S. (2018). A comparative phytochemical screening of four epidendroid orchids of Kerala, India. *Journal of Orchid Society of India*, *32*, 41–43.

Kalaiarasan, A., & John, S. A. (2011). Some bioactive constitutents of GC-MS analysis of *Bulbophyllum kaitense Rechib.* stem Eastern Ghats of India. *International Journal of Pharma and Bio Sciences*, *2*(4).

Kalaiarasan, A., Kumar, P., & John, S. A. (2012a). Antimicrobial activity of *Bulbophyllum kaitense* Rechib. stem of eastern peninsular flora in India. *Nature and Science*, *2*(10), 41–44.

Kalaiarasan, A., Kumar, P., & John, S. A. (2012b). Biochemical investigation of *Bulbophyllum* kaitense Rechib. root by GCMS Eastern Ghats of India. *Nature and Science*, *10*(2), 29–31.

Kumari, H., Nishteswar, K., Shukla, V. J., & Harisha, C. R. (2013). Development of pharmacognostic and phytochemical standards for pseudobulb of *Bulbophyllum neilgherrense*. *International Ayurvedic Medical Journal*, *1*(4), 1–8.

Kurzweil, H., & Lwin, S. (2014). *A guide to orchids of Myanmar*. Natural History Publications.

Lalitharani, S., Mohan, V. R., & Maruthupandian, A. (2011). Pharmacognostic investigations on Bulbophyllum albidum (Wight) Hook. f. *International Journal of PharmTech Research*, *3*(1), 556–562.

Leander, K., & Lüning, B. (1967). Studies on Orchidaceae alkaloids. VII. Structure of a Glucosidic alkaloid from *Malaxis congesta* comb. nov.(Rchb. f.). *Tetrahedron Letters*, *36*, 3477–3478.

Lin, D., Zhou, K., Hidayat, A., & Jin, X. H. (2020). *Bulbophyllum papuaense* (Orchidaceae), A new species from Indonesia. *PhytoKeys*, *138*, 125.

Lin, I. H., Chang, Y. S., Chen, I. S., & Hsieh, W. C. (2003). *The catologue of medicinal plant resources in Taiwan*. Committee on Chinese Medicine and Pharmacy, Department of Health.

Majumder, P. L., Kar, A., & Shoolery, J. N. (1985). Bulbophyllanthrin, a phenanthrene of the orchid *Bulbophyllum leopardium*. *Phytochemistry*, *24*(9), 2083–2087.

Majumder, P. L., Pal, S., & Majumder, S. (1999). Dimeric phenanthrenes from the orchid *Bulbophyllum reptans*. *Phytochemistry*, *50*(5), 891–897.

Majumder, P. L., & Sen, R. C. (1991). Bulbophyllanthrone, a phenanthraquinone from *Bulbophyllum odoratissimum*. *Phytochemistry*, *30*(6), 2092–2094.

Marasini, R., & Joshi, S. (2012). Antibacterial and antifungal activity of medicinal orchids growing in Nepal. *Journal of Nepal Chemical Society*, *29*, 104–109.

Nair, V. G., Prajapati, P. K., Nishteswar, K., Unnikrishnan, V., & Nariya, M. B. (2018). Analgesic and antiinflammatory activities of *Bulbophyllum neilgherrense* Wight. pseudobulb: A folklore plant. *Ayurveda*, *39*(2), 76

Nunes, E. L., Smidt, E. C., Stützel, T., & Coan, A. I. (2014). What do floral anatomy and micromorphology tell us about Neotropical *Bulbophyllum* section Didactyle (Orchidaceae: Bulbophyllinae). *Botanical Journal of the Linnean Society*, *175*(3), 438–452.

Pant, B., & Raskoti, B. B. (2013). *Medicinal Orchids of Nepal*. Himalayan Map House.

Prasad, G., Seal, T., Mao, A. A., Vijayan, D., & Lokho, A. (2021). Assessment of clonal fidelity and phytomedicinal potential in micropropagated plants of *Bulbophyllum odoratissimum*-An endangered medicinal orchid of Indo Burma megabiodiversity hotspot. *South African Journal of Botany*, *141*, 487–497.

Priya, K., & Krishnaveni, C. (2005). Antibacterial effect of *Bulbophyllum neilgherrense* wt. (Orchidaceae). An in vitro study. *Ancient Science of Life*, *25*(2), 50.

Qu, X. Y., Qin, S. Y., Yang, D. Q., Li, Q. S., & Peng, F. S. (2006). Study on resource and varieties of Guoshangye. *China Journal of Chinese Materia Medica*, 110–114.

Rahman, M., & Huda, M. K. (2021). Exploration of phytochemical, antioxidant and anti-inflammatory efficacy of the ethnomedicinal uses of ten orchids of Bangladesh. *Advancement in Medicinal Plant Research*, *9*(2), 30–39.

Tan, K. H., Tan, L. T., & Nishida, R. (2006). Floral phenylpropanoid cocktail and architecture of *Bulbophyllum vinaceum* orchid in attracting fruit flies for pollination. *Journal of Chemical Ecology*, *32*, 2429–2441.

Teoh, E. S. (2016). *Medicinal orchids of Asia* (Vol. 16). Springer.

Wu, B., He, S., & Pan, Y. J. (2006). New dihydrodibenzoxepins from *Bulbophyllum kwangtungense. Planta Medica*, *72*(13), 1244–1247.

Xu, J., Yu, H., Qing, C., Zhang, Y., Liu, Y., & Chen, Y. (2009). Two new *biphenanthrenes* with cytotoxic activity from *Bulbophyllum odoratissimum. Fitoterapia*, *80*(7), 381–384.

6 Phytochemicals and Therapeutic Importance of Some *Coelogyne* Lindl. (Orchidaceae)

An Overview

S.M. Khasim, J. Ramudu, M. Rahamtulla, and Venkatesh Rampilla

6.1 INTRODUCTION

The genus *Coelogyne,* having nearly 200 epiphytic taxa (Teoh, 2016), is distributed across India, China, Indonesia, and Fiji Islands with the main concentration in Borneo, Sumatra, and the Himalayas. They are dispersed in the lowland and montane forests. This genus has been studied extensively for its morphology and molecular characterization. However, very little work is done on the ethnomedicine of this genus. Apart from horticultural value, *Coelogyne* spp. have tremendous herbal/pharmaceutical potential. About six Chinese species have been used for medicinal purposes (Teoh, 2016). In India, *C. ovalis* is used as jeevanthi, which promotes healthier life. Janarthanan *et al.* (2020) studied the pharmacognostic standardization and phytochemical screening in the leaves of *C. breviscapa*. In this paper, an attempt has been made to review the ethnobotanical investigations on eight species of the genus *Coelogyne.*

6.1.1 *Coelogyne Barbata* Lindl. ex Griff.

Coelogyne barbata may grow as an epiphyte or lithophyte at an altitude of 1,000–1,800 meters. Psuedobulbs are clustered, carrying two stalked and 25–30 cm long thick leaves. It flowers in August-October. The inflorescence is an erect spike with white flowers and a contrasting blackish-brown lip at the center. Flowers are long-lasting, 7 cm across, and musk-scented. The lip is 3-lobed with its margin fringed and dark brown. It comes to flowering in August and is distributed in the Himalayas, Nepal, India, China, and Myanmar (Hawkes, 1965; Jin *et al.*, 2009).

6.1.1.1 Ethnobotany and Pharmacological Potentiality

In the Sichuan and Yunnan regions of China, the herb is collected, washed, and dried, then further sun-dried for storage. It is sweet in taste and cool in nature. It quenches thirst, stops coughs, and relieves pains. It is also used to treat sore throat, swelling of the scrotum, traumatic injuries, chappy extremities, and lung heat (Wu, 1994; Zhonghua Bencao, 2000). Some other therapeutic uses are summarized in Table 6.1 Zhonghua Bencao (2000).

6.1.1.2 Phytochemicals

The compounds such as batatasin -III, flavidin, flavidinin, and coelogin (Figure 6.1 A) were isolated from *C. barbata* (Veerraju *et al.*, 1989a, 1989b).

DOI: 10.1201/b23375-6

TABLE 6.1
External Application of *C. Barbata* (Decoction of 15–30 g of the Plant)

Therapeutic Use	Amount of Plant Content
Cough	Decoction with starchy banana 30 g
Throat (sore)	Decoction with starchy banana 30 g, 15 g of *Prunella*
Pain caused by hernia	Decoction of *C. barbata* along with 30 g starchy banana and 15 g Tangerine seed
Sprains	Fresh *C. barbata* paste, mixed with starchy banana paste, is applied to the injured area
Chapped hands and chapped foot	Equal quantity of *C. barbata* and starchy banana paste is mixed and applied to wounded portion

6.1.2 *C. Corymbosa* Lindl.

It is an epiphyte on the moist tree trunks and moist mountain tops at an elevation of 1,300–3,100 m. It is also terrestrial, having clustered pseudobulbs with a pair of oblong and coriaceous leaves (Figure 6.2A). The inflorescence is a raceme, curved or drooping, and bears two to four white or greenish-tinged flowers with distinctly attractive lips. In Bhutan, it comes to flowering from February to June (Gurong, 2006). In Nepal, it comes to flowering from March to May (Raskoti, 2009). *C. corymbosa* is distributed from Yunnan to Southern Tibet, including Myanmar, Sikkim, Bhutan, Nepal, and Northern India (Teoh, 2016). In India, it is found in Meghalaya (Khasia hills), Sikkim, and Arunachal Pradesh (Hegde, 1984).

6.1.2.1 Ethnobotany and Pharmacological Potential

The fresh juice of the pseudobulb is taken to heal burn wounds in India (Kumar, 2002; Hossain, 2011). The paste prepared from pseudobulbs is administered to the forehead to relieve pain (Pant and Raskoti, 2013; Bikarma Singh, 2022). In Yunnan and Xizang, pseudobulbs or whole plants have been administered to treat fractures.

6.1.2.2 Phytochemicals

Batatasin-III, flavidin, and Coelogin were reported for the first time from *C. corymbosa* (Veerraju *et al.*, 1989b). Further, 3(3-hydroxy phenyl)-5 methoxyphenol was also reported from this plant (Arya and Sahu, 2017). In addition to these, coeloginine, was also observed from the leaf extract (Bikarma Singh, 2022).

6.1.3 *C. Cristata* Lindl.

The *Coelogyne cristata* is an epiphyte on the moist tree trunks and branches. Pseudobulbs are oblong, and each carries two sessile, lanceolate leaves, 8–12 cm long. The inflorescence is long, containing 2–10 flowers. This plant has white, large, fragrant flowers. It comes to flowering from February to March in India and May in China (Teoh, 2016). It is distributed in Nepal and Northeastern India at an elevation of 1,000 to 2,000 m (Manandhar and Manandhar, 2002), Bangladesh, Myanmar, and Southern Tibet (Chen and Clayton, 2009). In India, it is found in the temperate Himalayas from Garhwal to Kumaon regions eastwards to Uttar Pradesh, Sikkim, West Bengal, Assam, Meghalaya, and Arunachal Pradesh (Pramanick, 2016).

6.1.3.1 Ethnobotany and Pharmacological Potential

The C. cristata has been used by Himalayan indigenous communities to treat fractured bones in animals (Jaiswal *et al.*, 2004). It is also used to treat dysentery and diarrhea in Myanmar (Naing *et al.*, 2010), whereas, in Nepal, the infusion of pseudobulbs is used to correcting constipation. Nepali communities also use this herb as an aphrodisiac (Pant and Raskoti, 2013). The fresh juice or paste of this plant has been used to relieve headaches, fever, and digestion (Subedi *et al.*, 2013). Folklore from the Kumaon region (Uttarakhand, India) says that it has been used to treat fractured bones (Sharma *et al.*, 2014). Native people in the Almora region apply the resin of pseudobulb externally on the injured portion of the body to coagulate the blood (Priti *et al.*, 2011). A poultice made from the whole plant is used as a fracture healing remedy in the folk tradition of Kumaon and Garhwal region, Uttarakhand Himalaya, India (Sharma *et al.*, 2014). Paste extracted from the whole plant without roots is applied to fractured parts of bones, whereas dry powder is given orally to treat bone injuries (Arya and Sahu, 2017). Lepcha tribal healers use the dried and finely powdered pseudobulbs of *C. cristata* along with a small amount of castor oil on foot cracks (Rahamtulla, 2022).

6.1.3.2 Phytochemicals

The compounds such as coelogin (Figure 6.1A), coeloginanthrene (Figure 6.1B), coeloginanthridin (Figure 6.1C), and coeloginin (Figure 6.1D) were reported from the whole plant of *C. cristata* (Majumder *et al.*, 1982a, 2001). Here, both coeloginanthrene and coeloginanthridin are 9, 10-dihydrophenanthrene derivatives. All these compounds are found to have biological activity of phytoalexins and endogenous plant hormones (Teoh, 2016).

6.1.3.3 Pharmacology

Mitra *et al.* (2018) studied the efficacy of the hydro-alcoholic extract of pseudobulbs against chronic fatigue syndrome (CFS) in aged Wistar rats. It is found to be non-toxic and significantly improved

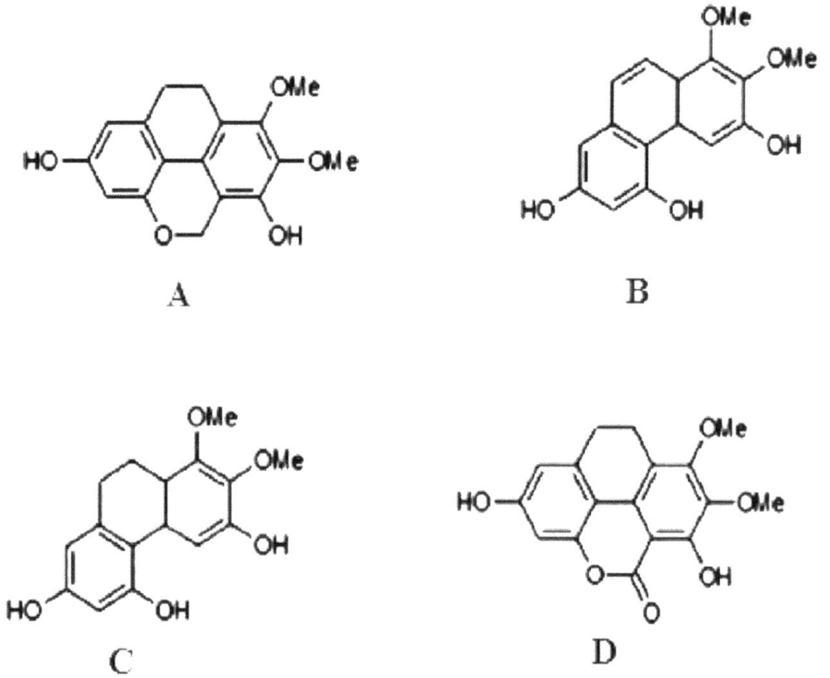

FIGURE 6.1 **A.** Coelogin, **B.** Coeloginanthrene, **C.** Coeloginanthridin and **D.** Coeloginin.

the locomotor movement in CFS-induced aged Wistar rats. Moreover, there is a marked diminution in lipid peroxidation, nitrite, and superoxide dismutase levels, and a significant enhancement of catalase levels in extract-treated rats (Mitra *et al.*, 2018). Prakash *et al.* (2021) observed that coelogin-treated Ovx mice promoted new bone formation, prevented the Ovx-induced deterioration of bone structure, and enhanced bone regeneration. Marasini and Joshi (2012) reported that the ethanolic extract of this plant is effective against the *Staphylococcus aureus* and moderately against the *Escherichia coli*. Further, pharmacological investigations of crude extract and isolated compounds from this species revealed rapid fracture healing properties and osteogenic potential (Sharma *et al.*, 2017).

6.1.4 *C. Flaccida* Lindl.

The *C. flaccida* is an epiphyte or saxicolous, consisting of spindle-shaped pseudobulbs with a coriaceous sheath. Leaves are leathery or narrowly lanceolate. Flowers are in a pendulous raceme, sparsely arranged, 2cm across, creamy white; the lip is dark brown. It flowers from February to May, lasting for 12–15 days. *Coelogyne flaccida* is found in Northeastern India, China, Laos, Myanmar, and Nepal (King and Pantling, 1898; Teoh, 2016).

6.1.4.1 Ethnobotany and Pharmacological Potential

The herb known as Guoshangye is a popular medicine in Guizhou and Yunnan provinces (Teoh, 2016). In China, the entire plant has been used to counter heat and dryness, promote the secretion of body fluids, and clear phlegm and cough (Wu, 1994; Zhonghua Bencao, 2000). Paste made from pseudobulbs is applied to the forehead to relieve the headache, whereas the juice is used to treat indigestion (Manandhar and Manandhar, 2002). The paste is also used to treat burns (Baral and Kurmi, 2006).

6.1.4.2 Phytochemicals

Phenanthrenes, stilbenoids, and flaccidin have been isolated from this herb (Majumder and Maiti, 1988). Isooxoflaccidin and soflaccidin have also been reported (Majumder and Maiti, 1991). Callosin originally reported from the *Agrostophyllum callosum* is also reported in *C. flaccida* (Majumder *et al.*, 1995; Kovacs *et al.*, 2007). Batatasin –III, imbricatin, and amoenomin were also recorded from this species (Veerraju *et al.*, 1989b).

6.1.5 *C. Nitida* (Wall Ex D.Don) Lindl.

An epiphytic orchid with oblong pseudobulbs, nearly 2cm long, carrying a pair of leathery lanceolate leaves. The raceme has six to eight white flowers, 4cm across, with their lip of 2cm long, tri-lobed.; the lateral lobeserrate (Figure 6.2 B). The flowering season is January to June. *Coelogyne nitida* is found in Northeastern India, Bhutan, Bangladesh, Nepal, Myanmar, Indo-China, and Thailand (Teoh, 2016).

Ethnobotany: Leaves of *C. nitida* are ground and made into a fine paste. A small amount of paste is applied externally to reset the bones and bone fractures (Rahamtulla *et al.*, 2020). Psuedobulb juice has been used in stomach-ache problems in Nepal (Manandhar and Manandhar, 2002; Baral and Kurmi, 2006). The paste and juice of pseudobulbs are also applied to headaches, fever, and burns (Subedi, 2011; Pant, 2013; Tsering *et al.*, 2017).

6.1.5.1 Phytochemicals

Chemical compounds such as ochrolide, a phenanthro-pyran, and ochron, a novel 9,10-dihydro-1,4 phenanthroquinone, along with coelonin, have been reported in *C. ochracea* (=*C. nitida*) (Bhaskar *et al.*, 1989, 1991). Another compound, ochrolic, a monomeric phenanthrene derivative and a precursor of phenanthrenes is also reported from this plant (Anuradha *et al.*, 1994). Besides, batatasin-III, flavidin, flavidinin, imbricatin, amoenomin, and coelogin were reported from this species (Veerraju *et al.*, 1989b). Further, *C. nitida* showed higher antioxidant activity due to the presence of a higher amount of quercetin content (Natta *et al.*, 2022).

FIGURE 6.2 *Coelogyne* species: **A.** *C. corymbosa*, **B.** *C. nitida*.

6.1.6 *C. Punctulata* Lindl.

C. punctulata is a perennial, epiphytic, or lithophytic orchid. It is distributed in the Himalayan region, Bangladesh, Myanmar, Thailand, and South-eastern China. Psuedobulbs are oblong and bright yellow when they are dried. Two leaves are found on the apex of the pseudobulb, and a papery sheath at the base. The leaf is lanceolate and petiolated. Inflorescence is 8–15cm long, with two to four white flowers. The lip is tri-lobed, and it has a central white keel.

6.1.6.1 Ethnobotany

Dried pseudobulb powder has been used to treat wounds and burns in Northern India (Das, 2004; Rahamtulla, 2022). It is also used to relieve the dry cough and bleeding resulting from trauma in Vietnam (Hung, 2014). Further, leaf extract has been used to induce vomiting in case of poisoning (Bikarma Singh, 2022).

6.1.6.2 Phytochemicals

Phytochemicals such as coeloginone, coelogin, quabain, and batatasin III have been reported from this plant (Bikarma Singh, 2022).

6.1.7 *C. Odoratissima* Lindl.

The C. odoratissima is a small-sized fragrant species growing on trees (epiphytic). Almost round pseudobulbs are clustered and enveloped by basal sheaths. These are pale green and heavily wrinkled and carry two leaves that are apical, membranous, elliptic-lanceolate, acute, plicate, and five-nerved petiolated. The terminal inflorescence is slender, drooping with two to three flowers. The flowers are white and fragrant. The lip has a small yellow patch in its center. It flowers from February to April. *C. odoratissima* is native to Southern India.

6.1.7.1 Ethnobotany

In South India, the paste of pseudobulbs has been used to treat skin diseases (Kumar *et al.*, 2015).

6.1.7.1.1 Phytochemicals

The chemical investigation clearly showed positive tests for phenolic compounds and saponins. Batatasin-III was recorded from this species (Veerraju *et al.*, 1989a).

6.1.8 *C. Stricta* (D.Don) Schltr.

The *C. stricta* is an epiphytic orchid with cylindrical to narrowly ovoid pseudobulbs, each carrying two lanceolate, leathery leaves marked with prominent veins, 18–30cm long. The inflorescence is erect, 60 cm long, with 4–10 flowers. It comes to flowering from March to June in Bhutan, rarely from October to November in Nepal (Gurong, 2006). *C. stricta* is found in Indochina, Yunnan, Myanmar, and Nepal (Chen and Clayton, 2009; Raskoti, 2009; Gurong, 2006). It is also found in the sub-tropical belt on tree trunks and rocky moss-covered hills in Arunachal Pradesh (Hegde, 1984).

6.1.8.1 Ethnobotany and Pharmacological Potential

The herb has been used to enhance bone healing and applied externally to fractured limbs in Northeast India (Trivedi *et al.*, 1980). A poultice prepared from pseudobulbs is administered to relieve headaches and fevers in Nepal (Baral and Kurmi, 2006; Pant and Raskoti, 2013). Pseudobulbs have been used to treat constipation and as an aphrodisiac; pseudobulb Juice is applied on wounds and boils, while pseudobulb gum is useful for sores (Pant, 2013; Rahamtulla, 2022).

6.1.8.2 Phytochemicals

A 9,10-dihydrophenanthrene known as 'coelogin' was isolated from *C. elata*, known as *C. stricta* (Majumder and Datta, 1984), and a 9,10 dihydrophenanthropyron has also been isolated from this plant (Majumder *et al.*,1982b).

6.1.8.3 Cytotoxic Potential

Thapa *et al.* (2020) assessed the anticancer potentiality of wild plants and *in vitro*-derived protocorms by using an MTT assay against the HeLa cancer cell line. The percentage of HeLa cell growth inhibition for wild plants extract is 41.99%, whereas *in vitro*-derived protocorm extract showed 20.78% at 400 μg/ml. The *in vitro* protocorms and plants are enriched in bioactive metabolites, which are likely to be useful for the commercial production of secondary metabolites that are useful to mankind. However, wild pseudobulb extracts showed more growth inhibition of cancer cells compared to *in vitro*-derived protocorms. The low toxic activity of protocorms must be due to the immaturity of protocorms, where more amounts of bioactive compounds could not be formed (Thapa *et al.*, 2020). Further, they observed that the wild plant extract showed a moderate effect toward cervical cancer cells.

6.2 CONCLUSION

From the literature, it is evident that the genus *Coelogyne* has immense potential in treating various diseases. There is a need to work on the isolation and purification of chemical compounds. Nowadays, most of the studies are mainly based on *in vitro* approaches only. Therefore, the investigation of animal models is essential to confirm the various therapeutic properties of the genus *Coelogyne*. Furthermore, because of their high medicinal potential and conservation value, these orchids require prompt bioprospecting.

6.2.1 Acknowledgments

The authors acknowledge Prof. D. Ramachandran, Chemistry Department, Acharya Nagarjuna University, for his generosity in drawing the chemical structures on ChemDraw software.

REFERENCES

Anuradha, V., Prakasa Rao, N.S., and Bhaskar, M.U. (1994). "Ochrolic acid, a precursor to phenanthrones from *Coelogyne ochracea*". Phytochemistry 36: 1515–1517.

Arya, K.R., and Sahu, N. (2017). "Ethnopharmacological developments in traditional bone healing plants: *Ulmus wallichiana*, *Coelogyne cristata* and *Pholidota articulata* for osteoporosis management". International Journal of Pharma Research. Health Science. 5(4): 1750–1754.

Baral, S.R., and Kurmi, P.P. (2006). "*A compendium of medicinal plants in Nepal*". Mrs. Rachana Sharma and IUCN, Kathmandu.

Bhaskar, M.U., Rao, J.M., and Rao, N.S.P. (1991). "Ochrone A, a novel 9,10-dihydro-1,4-phenanthroquinone from *Coelogyne ocracea*". Journal of Natural Products 54: 386–389.

Bhaskar, M.U., Rao, L.J.M., Rao, N.S.P., and Rao, P.R.M. (1989). "Ochrolide, a phenanthropyrone from *Coelogyne ochracea*". Phytochemistry 28(12): 3545–3546.

Bikarma Singh (2022). Therapeutic Himalayan herbs: Folklore uses, bioactive phytochemicals, and biological activities of medicinal orchids used by Nomads Indian. Journal of Natural Products and Resources 13(1): 94–104.

Chen, X.Q. and Clayton, D. (2009). *Coelogyne*. In: Chen, X.Q.; Zj, L.; Zhu, G.H. (Eds.) "Flora of India-Orchidaceae". Science Press Beijing. 315–325.

Das, S.P. (2004). Indian orchids in indigenous medicine system. In: Brirto, S.J. (Ed.) "*Orchid diversity and conservation*". The Rapinat Herbarium, St. John's College, Tiruchirappalli.

Gurong, D.B. (2006). "*An illustrated guide to the orchids of Bhutan*". DSB Publications, Thimpu.

Hawkes, A.D. (1965). "*Encyclopedia of cultivated orchids*". Faber & Faber, London.

Hegde, S.N. (1984). "*Orchids of Arunachal Pradesh*". Forest Department, Itanagar, Arunachal Pradesh, India.

Hossain, M.M. (2011). "Therapeutic Orchids: Traditional uses and recent advances-An overview". Fitoterapia, 82: 102–140.

Hung, T. (2014). Lan Thanh Dam- *Coelogyne*: duroc tinh (Thanh lan- *Coelogyne*: Pharmacology). www.hoalanvietnam.org

Jaiswal, S., Singh, S.V., and Singh, H.N. (2004). "Plant used for tissue healing of animals". Natural Product Radiance 3: 284–292.

Janarthanan, L., Karthikeyan, V., Senniappan, P., and Loganathan, J. (2020). Pharmacognostic standardization and phytochemical studies of *Coelogyne breviscapa* Lindl Leaves. Indian Journal of Natural Sciences 11(63): 28424–28431.

Jin, X.H., Zhao, X.D., and Shi, X.C. (2009). "*Native orchids from Gaoligongshan mountains China*". Science Press, Beijing.

King, G. and Pantling, R. (1898). "*The Orchids of the Sikkim-Himalaya Vol VIII*". Bishen Singh Mahendra Pal Singh, Dehra Dun.

Kovacs, A., Vasa, A., and Hohmann, J. (2007). "Natuaral phenanthrenes and their biological activity". Phytochemistry 69: 1084–1110.

Kumar, S. (2002). "The medicinal plants of North-East India". Scientific Publishers, Jodhpur.

Kumar, S.J., Ashok Kumar, R., Uma, G., Subbaiyan, B., Aravindhan, V., and Balasubramaniam, V. (2015). "Survey and documentation of commercially sold medicinal plants in local markets of Velliangiri Hills (Poondai), Coimbatore district, Tamil Nadu, India". International Journal of Recent Advances in Multidisciplinary Research 2(12): 1047–1055.

Majumder, P.L., Bandyopadhyay, D., and Joardar, S. (1982a). "Coelogin and coeloginin: Two novel 9,10-dihydrophenanthrene derivatives from the orchid *Coelogyne cristata*". Journal of Chemical Society 1: 1131–1136.

Majumder, P.L., Banerjee, S., Maiti, D.C., and Sen, S. (1995). "Stilbenoids from the orchids, *Agrostophyllum callosum* and *Coelogyne flaccida*". Phytochemistry 39: 649–653.

Majumder, P.L. and Datta, N. (1984). "Structure of Oxoflavidin, a 9,10-dihydrophenanthropyrone from *Coelogyne elata*". Phytochemistry, 23: 671–673.

Majumder, P.L., Laha, S., and Datta, N. (1982b). "Coelonin, a 9,10- dihydrophenanthrene from the orchids *Coelogyne ochracea* and *Coelogyne elata*". Phytochemistry 21: 478–480.

Majumder, P.L. and Maiti, D.C. (1988). "Flaccidin, a 9,10-dihydrophenanthropyran derivative from the orchids *Coelogyne flaccida*". Phytochemistry 27(3): 899–901.

Majumder, P.L. and Maiti, D.C. (1991). "Isoflaccidin and isooxoflaccidinin, stilbenoids from *Coelogyne flaccida*". Phytochemistry 30(3): 971–974.

Majumder, P.L., Sen, S., and Majumder, S. (2001) "Phenanthrene derivatives from the orchid *Coelogyne cristata*". Phytochemistry 58(4): 581–586

Manandhar, N.P. and Manandhar, S. (2002). "*Plants and people of Nepal*". Timber, Portland.

Marasini, R. and Joshi, S. (2012) "Antibacterial and antifungal activities of medicinal orchids growing in Nepal". Journal of Nepal Chemical Society 29: 104–109.

Mitra, A., Sur, T.K., and Hazra, J. (2018). "Effect of *Coelogyne cristata* Lindley in alleviation of chronic fatigue syndrome in aged Wistar rats". Journal of Ayurveda Integrative Medicine 9(4): 266–271.

Naing, A.H., Myint, K.T., and Hwang, Y.T. (2010). "Micropropagation and conservation of the wild medicinal orchids, *Coelogyne cristata*". Horticultural Environment Biotechnology. 51(2): 109–114.

Natta, S., Mondol, M.S.A., Pal, K., Mandal, S., Sahana, N., Pal, R., Pandit, G.K., Alam, B.K., Das, S.S., Biswas, S.S., and Kalaivanan, N.S. (2022). Chemical composition, antioxidant activity and bioactive constituents of six native endangered medicinal orchid species from north-eastern Himalayan region of India. South African Journal of Botany, 150: 248–259.

Pant, B. (2013). "Medicinal orchids and their uses: Tissue culture a potential alternative for conservation". African Journal of Plant Science 7: 448–467.

Pant, B. and Raskoti, B.B. (2013). "*Medicinal orchids of Nepal*". Himalayan Map House (P) Ltd., Kathmandu.

Prakash, R., Misra, T., Dev, K., Kuldeep, J., Ahmad John, A., Tripathi, A., Sharma, C., Arya, K.R., Kumar, B., Siddiqui, M.I., Tadigoppula, N. and Singh, D. (2021). Phenanthrenoid Coelogin isolated from *Coelogyne cristata* exerts osteoprotective effect through MAPK-Mitogen-activated protein Kinase signalling pathway. Calcified Tissue International 109: 32–43.

Pramanick, D.D. (2016). "Pharmacognostic studies on the pseudobulb of *Coelogyne cristata* Lindl. (Orchidaceace)-An epiphytic orchid of ethno-medicinal importance". Journal of Pharmacognosy and Phytochemitry 5(1): 120–123.

Priti, K., Joshi, G.C., and Tewari, L.M. (2011). "Diversity and status of ethno-medicinal plants of Almora District in Uttarakhand, India". Biodiversity and Conservation 3(7): 298–326.

Rahamtulla, M. (2022). Molecular and Ethnobotanical studies in some Indian Epidendroideae with special reference to Micropropagation of *Dendrobium aphyllum* (Roxb.) C.E.C. Fischer. Doctoral Thesis, Acharya Nagarjuna University, Guntur.

Rahamtulla, M., Rampilla, V., and Khasim, S.M. (2020). "Distribution and ethnomedicinal importance of orchids of Darjeeling Himalaya, India". Indian Forester 146(8): 715–721.

Raskoti, B.B. (2009). "*The orchids of Nepal*". Bhakta Bahadur Raskoti and Rite Ale, Kathmandu.

Sharma, C., Irshad, S., Khatoon, S. and Arya, K.R. (2017). Pharmacognostical evalution of Indian folk-traditional plants Coelogyne cristata and Pholidota articulate used for healing fractures. Indian Journal of Experimental Biology 55: 622–627.

Sharma, C., Mansoori, M.N., and Dixit, M.N. (2014). "Ethanolic extract of *Coelogyne cristata* Lindley (Orchidaceae) and its compound coelogin promote osteoprotective activity in overiectomized oestrogen deficient mice". Phytomedicine 21(12): 1702–1707.

Subedi, A. (2011). "*New species, pollinator interactions and pharmaceutical potential of Himalayan orchids*". Doctoral Thesis, Leiden University, The Netherlands.

Subedi, A., Kumar, B., and Choi, Y. (2013). "Collection and trade of wild-harvested orchids in Nepal". Journal of Ethnobiology and Ethnomedicine 9: 64–73.

Teoh, E.S. (2016). "*Medicinal orchids of Asia*". Springer Nature, Switzerland.

Thapa, B.B., Thakur, L.S., Joshi, P.R., Chand, K., Rajbahak, S., Sah, A.K., Shrestha, R., Paudel, M.R., Park, S.P., and Pant, V. (2020). "Ex situ conservation and cytotoxic activity assessment of native medicinal orchid: *Coelogyne stricta*". Journal of Plant Biotechnology 47: 330–336.

Trivedi, V.P., Dixit, R.S., and Lal, V.K. (1980). "Orchids in the drug markets of Bareilly, Knapur and nearby districts". Nagarjun (Calcutta) 23(8): 157–163.

Tsering, J., Tam, N., Tag, H., Gogoi, B.J., and Apang, O. (2017) "Medicinal orchids of Arunachal Pradesh: A review". Bulletin of Arunachal Forest Research 32 (1&2): 1–16.

Veerraju, P, Prakasa Rao, N.S., Jaganmohana Rao, L., Jagannadha Rao, K.V., and Mohana Rao, P.R. (1989a) "Amoenumin, a 9,10-dihydro-phenanthro-(4,5-b,c,d)-pyran from *Dendrobium amoenum*". Phytochemistry 28: 950–951.

Veerraju, P., Prakasa Rao, N.S., Jagan Mohana Rao, L., Jagannadha Rao, K., and Mohana Rao, P.R. (1989b) "Bibenzyls and phenanthroids of some species Orchidaceae". Phytochemistry 28: 3031–3034.

Wu, X.R. (1994) "*A concise edition of medicinal plants in China*". Guangdong Higher Education Publication House, Guangdong.1994.

Zhonghua Bencao (ed.). (2000) "Health department and National Chinese Management Office". Shanghai Science and Technology, Shanghai.

7 *Crepidium* Orchids and their Therapeutic Potential

Anu Prabha and Vikas Kushwaha

7.1 INTRODUCTION

The genus *Crepidium* contains about 300 species of orchids and approximately 250 broad-leaved terrestrial species. *Crepidium* is derived from the Greek word *krepidium*, which means "little boat," possibly referring to the shape of the labellum. In 1825, Carl Ludwig Blume officially described the genus *Crepidium*, and the description was published in his book Bijdragen tot de flora van Nederlandsch Indië. Ludwig (1825) and Szlachetko (1995) revised the very large genus *Malaxis*, and almost all Malesian species were transferred to the genera *Crepidium* or *Dienia* as a result of this revision (Pridgeon et al., 2001).

Crepidium species are distributed in tropical countries such as China, the Indian subcontinent, Southeast Asia, northern Australia, New Guinea, and various Pacific and Indian Ocean islands. The greatest diversity is found in New Guinea, where approximately ninety species exist. China is home to seventeen species, five of which are endemic (Zhao et al., 2009). In Australia, five endemic species are found in tropical Queensland and the Northern Territory. Many species grow on the rainforest floor, but some are epiphytes that grow on mossy tree trunks (Zhao et al., 2009).

Plants within this genus are typically small to medium in size, evergreen, and predominantly terrestrial, although a few species can grow as epiphytes. They feature two to many relatively large, pleated leaves, with their petioles encircling the stem, often lasting for up to three years. The flowers, which are typically small or tiny and do not resupinate, are located at the end of the flowering stem. These flowers exhibit a range of colors, including green, yellow, red, brown, or purple, and their sepals and petals tend to be spreading. The dorsal sepal is typically longer than the lateral sepals, while the petals are usually shorter and narrower than the sepals. The labellum is erect, flat, and typically undivided, often having two ear-like lobes near its base. The column is short and features two relatively large wings, and no nectar is produced. Additionally, two pairs of waxy, yellow pollinia are present (Teoh, 2021) (Figure 7.1).

Medicinal plants are important not only as traditional medicines but also as trading commodities. The role of information-derived ethnomedicine and its utility in drug discovery is critical. In Ayurveda, *Crepidium acuminatum* is classified as part of the "Ashtvarga" group of eight medicinal herbs, and its dried pseudobulbs are rich in bioactive compounds having antioxidant, cardioprotective, and anti-aging properties (Balkrishna et al., 2012). It's also a key ingredient in "Chyawanprash," an age-old health tonic sold by a variety of pharmaceutical companies around the world. The pharmacological and phytochemical analysis revealed that *Crepidium* is a rich source of secondary metabolites like alkaloids, glycosides, flavonoids, eugenol, stigmasterol, etc., and has curative properties like diuretic, anti-rheumatic, anti-inflammatory, anticarcinogenic, hypoglycemic, antimicrobial, anticonvulsive, neuro-protective and antiviral (Jalal et al., 2008). The purpose of this chapter is to deliberate about the application of *Crepidium* orchids in traditional and current systems of medicine.

DOI: 10.1201/b23375-7

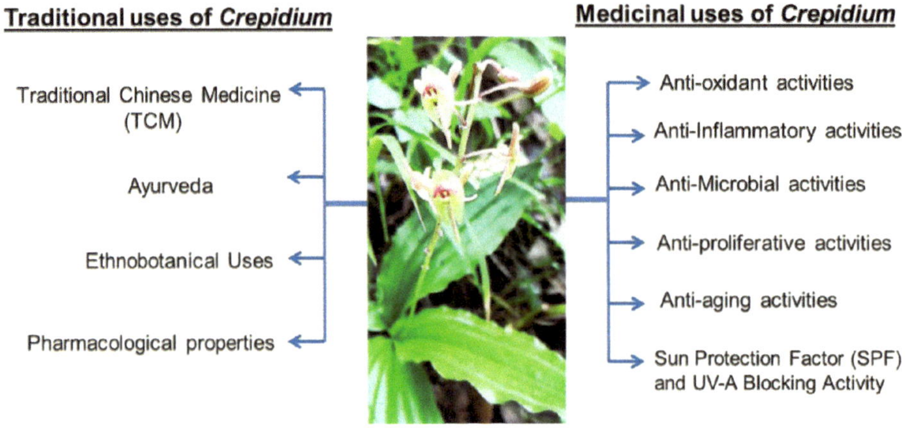

FIGURE 7.1 Traditional and medicinal uses of *Crepidium* spp.

7.2 TRADITIONAL USES OF *CREPIDIUM*

Crepidium has been traditionally used in various systems of traditional medicine, including Ayurveda and Traditional Chinese Medicine (Figure 7.1). While it may not be widely recognized or extensively studied in the context of modern Western medicine, some of its potential applications and uses in the current system of medicine include:

1. **Traditional Chinese Medicine (TCM):** Some species of *Crepidium* have been used in TCM for their potential medicinal properties. They may be incorporated into herbal formulations to address specific health issues, although their use is not as common or well-documented as other herbs in TCM (Gupta et al., 2015).
2. **Ayurveda:** In Ayurveda, *Crepidium* species may be employed in herbal remedies, especially for conditions related to reproductive health, sexual vitality, and certain gastrointestinal disorders. It's important to note that the use of specific *Crepidium* species and their applications can vary in Ayurvedic practice (Pushpa et al., 2011).
3. **Ethnobotanical Uses:** In various indigenous and traditional healing systems, specific species of *Crepidium* may have local uses for treating ailments like digestive problems, skin issues, or as a general tonic. These uses may not be well-documented in scientific literature but are part of traditional knowledge. *Malaxis acuminata* has a long history of utilization, dating back to ancient times. It has been incorporated into the preparation of Chyawanprash. Meanwhile, *Malaxis muscifera* is a member of the Asthavarga group of herbs, and its formulations are employed as tonics and rejuvenating agents. Both plants are employed to enhance semen quantity and stimulate semen production. Their pseudobulbs possess properties such as sweetness, cooling effects, aphrodisiac qualities, astringency, anti-dysentery attributes, fever-reducing abilities, and tonic properties. They find application in various conditions like infertility, imbalances in pitta and vata, seminal weakness, both internal and external hemorrhages, dysentery, fever, emaciation, burning sensations, and overall debility. An external paste made from the pseudobulb is suitable for treating insect bites, and when combined with other herbs, it is used in rheumatism treatment. According to Ayurvedic principles, these plants have a sweet taste, a cooling nature, pacify vata, and exacerbate kapha. They have also been employed in pharmacology to manage bleeding diathesis, burning sensations, and phthisis (lung disease) (Pal et al., 2022).

4. **Pharmacological properties:** Some scientific research may be conducted to explore the potential pharmacological properties of *Crepidium* species. This research can lead to a better understanding of their potential benefits and applications in modern medicine. However, it's essential to be cautious and ensure that any medicinal use is supported by scientific evidence.

It's important to note that the use of *Crepidium* in modern medicine is not as widespread or well-established as that of more commonly studied and recognized medicinal plants. Additionally, the safety, efficacy, and dosage recommendations for specific *Crepidium* species are not well-documented.

7.3 BIOACTIVE COMPOUNDS ISOLATED FROM *CREPIDIUM*

7.3.1 Phenolic Compounds

Phenolic compounds, a major plant secondary metabolite class, aid plant defense against various threats, including pathogens, diseases, parasites, and predators. They also participate in physiological processes like antioxidation and lipid peroxidation stabilization (Bhattacharyya et al., 2014). A phytochemical examination of methanolic extracts from the pseudobulbs, leaves, and stems of *C. acuminatum* revealed their abundance of flavonoids, phenolic acids, and tannins. Among the in vitro-derived plants, the methanolic leaf extract displayed the highest concentrations of total phenolics (44.34 ± 0.7 mg GAE/g DW) and flavonoids (24.56 ± 0.5 mg QE/g DW), whereas the aqueous stem extracts from wild plants exhibited the lowest levels of phenolic compounds (7.82 ± 0.2 mg GAE/g DW) and flavonoids (3.78 ± 0.1 mg QE/g DW) (Bose et al., 2017). Analysis using HPLC-ESI-QTOF-MS/MS identified Catechin, phloridzin, and rutin as the flavonoid compounds present in *C. acuminatum*, while the phenolic acids included Caffeic acid, chlorogenic acid, ellagic acid, 3-hydroxy benzoic acid, 4-hydroxy benzoic acid, protocatechuic acid, 3-hydroxy cinnamic acid, and p-coumaric acid (Figures 7.2 and 7.3).

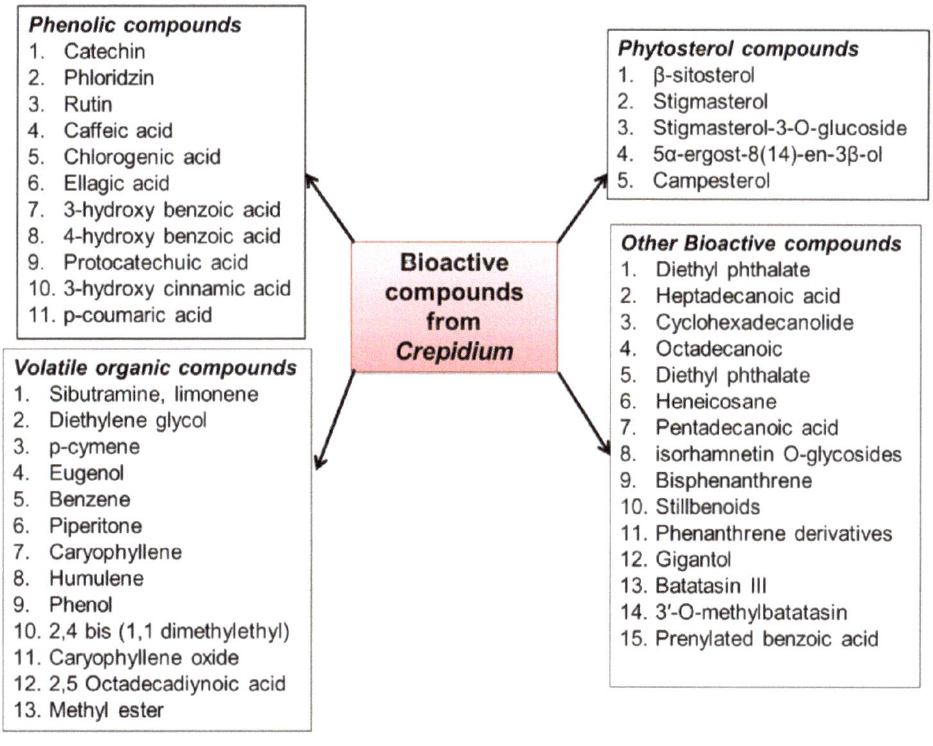

FIGURE 7.2 Bioactive compounds present in *Crepidium* spp.

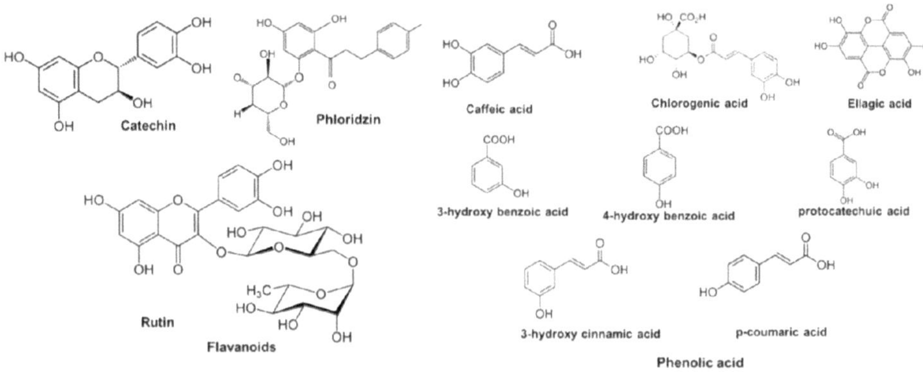

FIGURE 7.3 Phenolic compounds in *Crepidium* spp.

7.3.2 Phytosterol

Plant sterols are an essential component of the membranes of all eukaryotic organisms that control membrane fluidity and permeability, although some plant sterols have a specific function in signal transduction (Valitova et al., 2016). Phytosterols such as β-sitosterol and stigmasterol bear a structural resemblance to cholesterol and are present in various plant species, including orchids (Aboobucker and Suza, 2019). In *C. acuminatum*, the two primary sterol compounds identified are stigmasterol and β-sitosterol. β-sitosterol is a well-known phytosterol with recognized therapeutic properties, including anticancer, estrogenic, antidiabetic, and anti-aging effects.

In 2017, Singh and colleagues used Nuclear Magnetic Resonance (NMR) to detect and characterize three compounds, namely β-sitosterol, stigmasterol, and stigmasterol-3-O-glucoside, in the ethyl acetate fraction of *C. acuminatum* methanolic extract (Singh et al., 2017). According to Bhattacharyya et al. in 2022, the leaf tissue and pseudobulb of micropropagated *M. acuminata* demonstrated higher concentrations of both β-sitosterol (39.69 μg/100mg) in leaf and (36.25 μg/100mg) in pseudobulb. The pseudobulb is traditionally employed in herbal preparations (Bhattacharyya et al., 2022b).

GC-MS analysis of the floral parts of *C. hoi, C. acutangulum*, and *C. luniferum* revealed their significant sterol content, including stigmasterol and β-sitosterol (ranging from 1.03% to 4.66%). Additionally, C. hoi was found to contain trace amounts of 5α-ergots-8(14)-en-3β-ol (0.86%) and campesterol (3.62%) (Margońska et al., 2021) (Figure 7.4).

7.3.3 Volatile Organic Compounds

Plant volatile compounds are a complex mixture of low molecular weight lipophilic molecules that have low melting points (Pichersky, 2006). Volatile compounds are categorized into three groups: terpenoids, phenylpropanoids/benzenoids, and fatty acid derivatives. GC-MS analysis of a methanolic extract of the leaf and stem of *C. acuminatum* showed the presence of various volatile compounds such as sibutramine, limonene, diethylene glycol, p-cymene, eugenol, benzene, and piperitone (Bose et al., 2017).

The lipid composition of floral extract of *C. hoi* comprises saturated and unsaturated hydrocarbons (46%), followed by substantial levels of free glycerol and aliphatic alcohols (33%), sterols (13%), with fatty acids making up a minor fraction (3%). A notably higher proportion of fatty acids was also detected in *C. acutangulum* (46.61%) and *C. luniferum* (55.35%) (Margońska et al., 2021). The methanolic extracts of *Crepidium acuminatum* pseudobulb contains mainly essential oils which

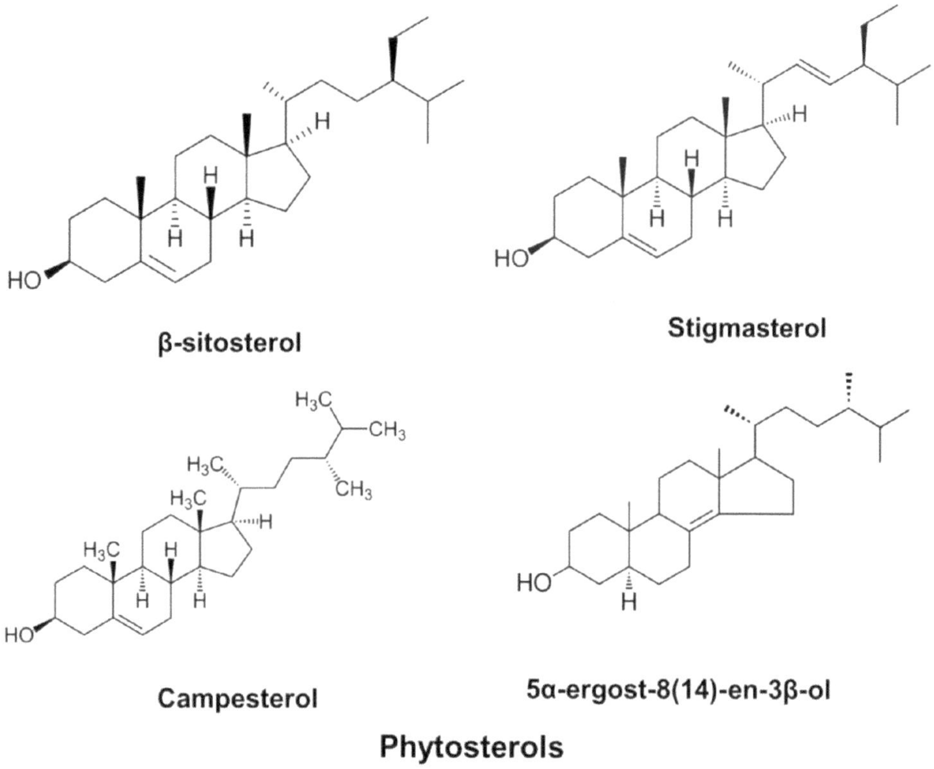

FIGURE 7.4 Phytosterols compounds in *Crepidium* spp.

are caryophyllene 74.89%; Eugenol 5.56%; Humulene 5.48%; phenol, 2,4 bis (1,1 dimethylethyl) 5.57%; Caryophyllene oxide 5.33%; 2,5 Octadecadiynoic acid, methyl ester 3.16% (Arora et al., 2017) (Figure 7.5).

7.3.4 Other Bioactive Compounds

Besides the previously-mentioned bioactive compounds, GC-MS analysis of methanol/water extract of root of *M. acuminata* yielded 67 bioactive compounds including heptadecanoic acid, methyl ester, cyclohexadecanolide, octadecanoic, diethyl phthalate, heneicosane, hexadecane, docosane, pentadecanoic acid possess various biological activities (Raval et al., 2016). HPLC-ESI-QTOF-MS analysis of ethyl acetate fraction of *M. acuminata* possessed flavonoids (isorhamnetin O-glycosides; isomer 1 & 2), bisphenanthrene, stillbenoids, phenanthrene derivatives (lusianthrin and 2,3-dimethoxy-9,10-dihydrophenanthrene-4,7-diol), gigantol, batatasin III, and 3′-O-methylbatatasin, liparacid C, and prenylated benzoic acid (Singh et al., 2017). GC–MS analysis of methanolic leaf and stem extracts in the wild and in vitro-derived plants of *M. acuminata* also contains glycerol, ribitol, and myo-inositol.

7.4 APPLICATIONS OF *CREPIDIUM* IN CURRENT SYSTEM OF MEDICINE

Phytoconstituents present in the bioactive extracts or fractions of root, stem, leaf, and pseudobulb of *C. acuminatum* evaluated for their biological activity.

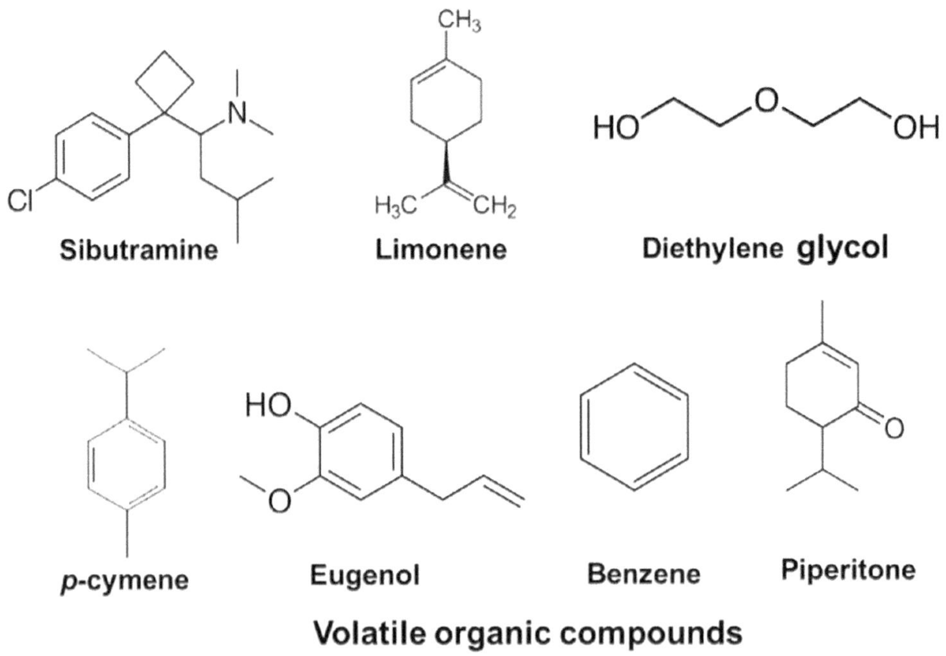

FIGURE 7.5 Volatile organic compounds in *Crepidium* spp.

7.4.1 Antioxidant Activities

The term "antioxidant activity of plant extracts" pertains to the ability of bioactive constituents obtained from different plant parts, such as leaves, stems, roots, or pseudobulbs, to mitigate oxidative stress within living organisms and prevent the oxidative deterioration of essential biomolecules within human tissues and cells. The overproduction of reactive oxygen species (ROS) or free radicals is a prominent factor in the development of various chronic diseases, including Alzheimer's disease, various types of cancer, and diabetes, which is counteracted by the presence of antioxidants (Natividad et al., 2020). Several methods have been developed and utilized to evaluate the antioxidant activities *in C. accuminata*, including in vitro assays for ferric-reducing antioxidant potential assay (FRAP), superoxide scavenging assays, free radical scavenging and reducing assays such as 2,2-azinobis (3-ethylbenzoline-6-sulfonic acid) radical assay (ABTS) and 1-diphenyl-2-picrylhydrazylfree radical-scavenging assay (DPPH). 1, 1-diphenyl-2-picryl hydrazyl (DPPH) radical scavenging activity, Reduction capability by Fe_3+ to Fe_2+ transformation method and Hydrogen peroxide scavenging method revealed that *Malaxis acuminatae* extract showed a good antioxidant activity (Sharma et al., 2011). Giri and colleagues showed that the pseudobulbs of extract of the M. accumulata have strong antioxidant potential in ABTS assay (4.02 mM AAE/100 g DW), DPPH assay (1.10 mM AAE/100 g DW), and FRAP assay (1.18 mM AAE/100 g DW) (Giri et al., 2017). Bose et al. 2017 demonstrated that methanolic leaf extract from in vitro-derived plants of *M. accuminata* showed significant antioxidant activity 42.66 ± 1.8 μg/ml (IC50). However, stem extracts showed moderate DPPH activity, where the lowest radical scavenging potential was observed in aqueous stem extracts of wild plants (178.56 ± 2.9 μg/ml). Methanolic leaf extracts of tissue-cultured plants showed significant ABTS•+ radical scavenging activity [13.32 ± 1.1 μg/ml (IC50)] (Bose et al., 2017). The analysis of *Malaxis acuminta* D. Don (Jeevak) extract by GC/MS revealed the presence of several compounds, including methyl ester, 1-Butanol, 3-methyl-, acetate, octadecanoic acid, butyl oleate, eicosanoic acid, octadec-9-enoic acid, thiirane, octyl-9,9-dimethoxybicyclo[3.3.1]nona-2,4-dione, methyl (6E)-6-octadecenoate, DBP; 1,2-benzenedicarboxylic acid, bis(2-methylpropyl) ester, Cerasynt Special, Univol U 316S, and oxirane, tetradecyl. These compounds are known to possess antioxidant properties (Raval et al., 2016).

7.4.2 Anti-Inflammatory Activity

Inflammation serves as the body's innate protective response to infection or other detrimental stimuli. Prolonged or excessive inflammation can have adverse effects and is linked to numerous health issues, including autoimmune diseases, arthritis, and persistent inflammatory ailments. Anti-inflammatory agents work to diminish bodily inflammation by regulating the release of various proinflammatory mediators, such as cytokines and chemokines (Turner et al., 2014). Plants have natural antioxidants and enzyme inhibitors in the form of bioactive compounds that serve diverse functions and possess anti-inflammatory actions (Virshette et al., 2019).

The ethanolic extract from pseudobulbs has been documented for its in vivo pain-relieving and anti-inflammatory effects (Rath et al., 2011). Additionally, the GC/MS analysis of the methanolic extract obtained from dried pseudobulbs of *Malaxis acuminata* D.Don revealed the presence of compounds such as Pentadecanoic acid, 6-Octadecenoic acid methyl ester (Z-isomer), and Eicosanoic acid methyl ester, and eugenol that possess anti-inflammatory activity (Raval et al., 2016). A study performed by Bose et al. (2017) demonstrated that methanolic leaf extracts (IC50: 14.32 μg/ml) and stem extracts (IC50: 16.2 μg/ml) from tissue culture-derived of *M. accunimata* plants showed noteworthy anti-5-lipoxygenase activity. Although their potency was not as strong as that of a known 5-LOX inhibitor (IC50: 2.56 μg/ml), they exhibited the most robust anti-hyaluronidase activity with an IC50 of 60.36 μg/ml when compared to oleanolic acid, which had an IC50 of 32.45 μg/ml (Bose et al., 2017).

7.4.3 Antimicrobial Activity

Antimicrobial activity refers to the ability of a substance or agent to inhibit or kill disease-causing microbes such as bacteria, viruses, fungi, and other pathogens, thereby preventing their growth or spread. They include Antiseptics, Antibiotics, Antivirals, Antifungals, and Antiparasitics (Di Martino, 2022). Plant-derived natural compounds are highly promising candidates for serving as alternative sources of antimicrobial agents. Butanol extracts of *M. accuminata* showed significant antifungal (against Candida albicans) and antibacterial activities of against *Escherichia coli, Klebsiella aerogenes, Pseudomonas aeruginosa, Proteus mirabilis, Staphylococcus aureus* (Pushpa et al., 2011). GC/MS analysis of fractions 2, 10, and 12 of the methanolic extract of M. acuminata revealed the presence of various compounds, including Thiirane, octyl-, Tert-Hexadecanethiol, Methyl tetradecanoate, Heneicosane, Margaric acid, Methyl (6E)-6-octadecenoate, Hexadecanoic acid methyl ester, and E-8-Hexadecen-1-ol acetate, which have shown antimicrobial activity (Raval et al., 2016). Further, Different plant extracts such as hexane, chloroform, ethyl acetate, ethanol, and residual aqueous of pseudobulbs of *M. acuminate/C. acuminatum* exhibited potential antibacterial activity against Staphylococcus aureus, Bacillus subtilis, *Escherichia coli,* and *Pseudomonas aeruginosa* (Arora et al., 2017).

7.4.4 Anti-Proliferative Activity

Anti-proliferative activity is the ability of compounds to reduce or inhibit cell growth below the control through regulation of gene expression and initiation of signal transduction pathways and transcriptional activation of JAK-STAT (Xin et al., 2020). Crude ethanol extract of M. accuminata and its fractions hexane, ethyl acetate, butanol, water were assessed for anti-proliferative activity by Sulphorhodamine B (SRB) assay against four human different cancer cell lines such as A549 (non-small cell lung cancer cells), DU145 (human prostate carcinoma), DLD1 (human colorectal adenocarcinoma), and MCF-7 (human breast adenocarcinoma). The ethyl acetate fraction showed a significant potent anti-proliferative activity against cancer cell lines, such as A549 (70.29 ± 7.22), DLD1 (73.12 ± 8.58), MCF-7 (79.10 ± 9.62), and DU145 (68.65 ± 8.10) (Singh et al., 2017).

7.4.5 Anti-Aging Activity

The process of skin aging is intricate, affecting every layer of the epidermis and dermis, with the aim of fortifying and safeguarding skin stem cells. Plant stem cells possess anti-aging attributes, encouraging fibroblasts to produce collagen and promote skin rejuvenation (Miastkowska and Sikora, 2018). Orchid extracts from *Vanda coerulea* and *V. teres* combat aging by countering reactive oxygen species (ROS). They enhance mitochondrial function in keratinocytes by stimulating cytochrome c oxidase, while their natural mucilage effectively moisturizes and softens the skin (Hadi et al., 2015). Because of the antioxidant and anti-aging attributes found in the extract of Pseudobulbs from *M. acuminata*, it is employed in crafting formulations aimed at enhancing vitality and rejuvenating the skin (Bhattacharyya et al., 2022a). Collagen is a vital protein that serves as the structural foundation of our body and maintains the strength, elasticity, and integrity of various tissues, including skin, tendons, ligaments, and bones (Yifeng, 2023). Collagen and elastin are fundamental for maintaining the skin's structural integrity and elasticity. The methanolic leaf extract exhibited notable anti-collagenase activity with an IC50 value of 32.52 µg/ml. Moreover, it demonstrated a highly potent elastase inhibitory effect, with an IC50 of 32.24 µg/ml. As such, M. *acuminata* holds promise as a potential agent for skin rejuvenation and combating the signs of aging (Bose et al., 2017). Recent findings have reported that vandaterosides, the compounds responsible for the pink and purple coloration of orchids, possess anti-aging properties for the skin (Begum, 2022).

7.4.6 Sun Protection Factor (SPF) and UV-A Blocking Activity

Both short-term and prolonged exposure to detrimental radiations, such as sunlight and ultraviolet radiation, result in a range of skin alterations by causing oxidative damage to the lipids, proteins, and DNA on the outer layer of the body (Sharma et al., 2020). UV-A (400–315 nm) and UV-B (315–280 nm) radiations cause the melanin pigments in the skin to oxidize (darken), creating erythema (sunburn). Plants contain diverse natural compounds with varying UV absorption abilities, driving toward SPF formulation. Some plant phenolics tested in vitro were mutagenic, possibly due to their pro-oxidant properties. Considering their human use in sunscreens, it's crucial to assess the extracts for toxicity (Srinivasan et al., 2018). The methanolic extracts from both wild and in vitro-cultivated *M. acuminata* plants exhibited significant UV-A blocking activity. Specifically, leaf and stem extracts, when used at a concentration of 250 µg/ml, demonstrated in vitro SPF values of 21.68 and 27.64 µg/ml, respectively. These outcomes are likely associated with the presence of numerous polyphenolic and flavonoid compounds found in the plant extracts (Bose et al., 2017).

7.5 CONCLUSION AND FUTURE REMARKS

Orchids are famous for their beautiful blooms, but their potential therapeutic properties are not widely recognized. Although various parts of the plant have yielded chemicals with therapeutic potential, clinical trials involving orchid components are scarce. A greater focus on clinical research could unlock new avenues for herbal remedies. *Crepidium* is valued for its rejuvenating properties in traditional Ayurvedic formulas and has been studied for its potential anti-aging, antioxidant, anti-inflammatory, sun protection, UV-A blocking, anti-proliferative, and antimicrobial effects. Further research is needed to understand the pharmacological benefits of the compounds extracted from *Crepidium* using various solvents and fractions, supporting their significance in ethnopharmacology. Identifying the active molecule's precise physiological mechanism of action is essential.

Additionally, establishing statistical differences among bioactive phytochemicals is crucial during drug development testing.

Now, high throughput sequencing, encompassing whole-genome and transcriptome sequencing technologies, has emerged as a reliable approach to efficiently generate extensive datasets of gene expression in a cost-effective and time-efficient manner. This has proven to be an invaluable resource for the development of markers tailored to specific traits in the context of molecular breeding, genetic diversity analysis, or population genetics research. This wealth of information serves as a valuable resource for expeditious genotyping, advancing molecular breeding efforts, and pinpointing functional markers to enhance specific traits.

In conclusion, *C. acuminatum/M. acuminata* exhibits vitality-enhancing and rejuvenating properties, supported by various pharmacological studies. Nevertheless, most of these phytochemical and pharmacological investigations are in their early stages and require more comprehensive scientific evidence.

REFERENCES

Aboobucker, S. & Suza, W. 2019. Why do plants convert sitosterol to stigmasterol? *Frontiers in Plant Science* 10:354.

Arora, M., Mahajan, A. & Sembi, J. 2017. Essential oils analysis of pseudobulbs of *Crepidium* acuminatum (D.DON) SZLACH by GC-MS. *Asian Pacific Journal of Health Science* 4:198–204.

Balkrishna, A., Srivastava, A., Mishra, R., Patel, S., Vashistha, R., Singh, D., Jadon, V. & Saxena, P. 2012. Astavarga plants-threatened medicinal herbs of the North-West Himalaya. *International Journal of Medicinal and Aromatic Plants* 2:661–676.

Begum, S. 2022. DNA of Singapore's national flower reveals an anti-ageing compound in orchid. *The Star*, 1 November (www.thestar.com.my/aseanplus/aseanplus-news/2022/11/01/dna-of-singapores-national-flower-reveals-an-anti-ageing-compound-in-orchid).

Bhattacharyya, P., Kumaria, S., Diengdoh, R. & Tandon, P. 2014. Genetic stability and phytochemical analysis of the *in vitro* regenerated plants of *Dendrobium nobile* Lindl., an endangered medicinal orchid. *Metagene* 2:489–504.

Bhattacharyya, P., Lalthafamkimi, L., Sharma, T., Kumar, D., Kumar, A. & Kumar, S. 2022a. Metabolic and tissue-specific expression profiling in micropropagated plants of *Malaxis acuminata*: An endangered medicinal orchid. *Plant Cell, Tissue and Organ Culture* 151:535–549.

Bhattacharyya, P., Sharma, T., Yadav, A., Lalthafamkimi, L., Deshwal, R., Swarnkar, M., Joshi, R., Shankar, R. & Kumar, S. 2022b. De novo transcriptome based insights into secondary metabolite biosynthesis in *Malaxis acuminata* (Jeevak) – A therapeutically important orchid. *Frontier in Plant Sciences* 13:954467.

Bose, B., Choudhury, H., Tandon, P. & Kumaria, S. 2017. Studies on secondary metabolite profiling, anti-inflammatory potential, in vitro photoprotective and skin-aging related enzyme inhibitory activities of *Malaxis acuminata*, a threatened orchid of nutraceutical importance. *Journal of Photochemistry and Photobiology B* 173:686–695.

Di Martino, P. 2022. Antimicrobial agents and microbial ecology. *AIMS Microbiolology* 8:1–4.

Giri, L., Belwal, T., Bahukhandi, A., Suyal, R., Bhatt, I., Rawal, R. & Nandi, S. 2017. Oxidative DNA damage protective activity and antioxidant potential of Ashtvarga species growing in the Indian Himalayan Region. *Industrial Crops and Products* 102:173–179.

Gupta, A., Mishra, R. & Bhati, M. 2015. Traditional medicinal uses, phytochemical profile and pharmacological activities of *Crepidium* acuminatum (D. Don) Szlach. *Indian Journal of Ancient Medicine and Yoga* 8:179–183.

Hadi, H., Razali, S. & Awadh, A. 2015. A comprehensive review of the cosmeceutical benefits of *Vanda* species (Orchidaceae). *Natural Product Communications* 10:1483–1488.

Jalal, J., Kumar, P. & Pangtey, Y, 2008. Ethnomedicinal orchids of Uttarakhand, Western Himalaya. *Ethnobotanical Leaflets* 12:1227–1230.

Ludwig, B. C. 1825. Bijdragen tot de flora van Nederlandsch Indië. *Batavia*, p. 387 (www.biodiversitylibrary.org/item/9224#page/105/mode/1up).

Margońska, H., Kozieradzka-Kiszkurno, M., Brzezicka, E., Haliński, Ł., Davies, K. & Lipińska, M. 2021. *Crepidium* sect. *Crepidium* (Orchidaceae, Malaxidinae)—Chemical and morphological study of flower structures in the context of pollination processes. *Plants* 10:2373.

Miastkowska, M. & Sikora, E. 2018. Anti-aging properties of plant stem cell extracts. *Cosmetics* 5:55.
Natividad, C., Santiago, A. & Alías Gallego, J. 2020. Quantification of the antioxidant activity of plant extracts: Analysis of sensitivity and hierarchization based on the method used. *Antioxidants* 9:76.
Pal, R., Meena, N., Dayamma, M. & Singh, D. R. 2022. Ethnobotany and recent advances in Indian medicinal orchids. In: *Orchids Phytochemistry, Biology and Horticulture* (pp. 1–27). Springer, Cham: Springer Nature.
Pichersky, N. D. 2006. *Biology of Floral Scent*. Boca Raton: CRC Press.
Pridgeon, A. M., Cribb, P. J., Chase, M. W. & Rasmussen, F. N. 2001. *Genera Orchidacearum* (Vol. 2). Oxford, UK: Oxford University Press.
Pushpa, M. M., Mahajan, N., Pankaj, G., Gurkirpal, S., Sumit, D. & Sakshi, S. 2011. *Malaxis acuminata*: Review. *International Journal of Research in Ayurveda and Pharmacy* 2:422–425.
Rath, C., Kumari, S., Bishnupriya, D., Mohanty, R. C., Renu, D., Padhi, M. & Ramesh, B. 2011. Phytopharmacognostical studies of two endangered species of *Malaxis* (Jeevak and rishibhak). *Pharmacognosy Journal* 3:77–85.
Raval, S., Pradipkumar, V., Mandavia, M. & Golakiya, B. 2016. Phytochemical analysis of *Malaxis* acuminta D. Don (Jeevak), an ingredient of Jeevaniya Group of Ashtavarga. *Indian Journal of Agricultural Biochemistry* 29:155.
Sharma, P., Garg, P., Dadhwal, S., Singh, G., Sharma, D. K. & Sharma, S. 2011. Antimicrobial activity of butanol extract of *Malaxis acuminata*. *Journal of Pharmacy Research* 4(8):2703–2704.
Sharma, T., Tyagi, V. & Bansal, M. 2020. Determination of sun protection factor of vegetable and fruit extracts using UV–Visible spectroscopy: A green approach. *Sustainable Chemistry and Pharmacy* 18:100347.
Singh, D., Kumar, S., Pandey, R., Hasanain, M., Sarkar, J. & Kumar, B. 2017. Bioguided chemical characterization of the antiproliferative fraction of edible pseudo bulbs of *Malaxis acuminata* D. Don by HPLC-ESI-QTOF-MS. *Medicinal Chemistry Research* 26:3307–3314.
Srinivasan, P., Inala, M., Hs, N., Avm, K. & Pamidimukkala, K. 2018. A pilot study on sun protection factor of plant extracts: An observational study. *Asian Journal of Pharmaceutical and Clinical Research* 11:67–71.
Szlachetko, D. L. 1995. Systema Orchidalium. *Fragmenta Floristica et Geobotanica Supplementum* 3:1–152.
Teoh, E. S. 2021. *Orchid Species from Himalaya and Southeast Asia*. Cham: Springer International Publishing.
Turner, M. D., Nedjai, B., Hurst, T. & Pennington, D. J. 2014. Cytokines and chemokines: At the crossroads of cell signalling and inflammatory disease. *Biochimica et Biophysica Acta (BBA)—Molecular Cell Research* 1843:2563–2582.
Valitova, J. N., Sulkarnayeva, A. G. & Minibayeva, F. V. 2016. Plant sterols: Diversity, biosynthesis, and physiological functions. *Biochemistry Biokhimiia* 81:819–834.
Virshette, S. J., Patil, M. K. & Somkuwar, A. P. 2019. A review on medicinal plants used as anti-inflammatory agents. *Journal of Pharmacognosy and Phytochemistry* 8:1641–1646.
Xin, P., Xu, X., Chengjie, D., Liu, S., Wang, Y., Zhou, X., Ma, H., Wei, D. & Sun, S. 2020. The role of JAK/STAT signaling pathway and its inhibitors in diseases. *International Immunopharmacology* 80:106210.
Yifeng, F. 2023. Collagen: The structural protein that supports body's strength and flexibility. *Journal of Food Technology and Preservation* 7(4):177.
Zhao, I. S., Chen, X., Chen, S. & Wood, J. J. 2009. *Crepidium* Blume, bijder. 387. 1825. *Flora of China* 25:229–234.

8 *Cypripedium* Orchids and their Utilization in Traditional and Modern Medicine

Anu Prabha, Vikas Kushwaha, Anamika Kumari, and Promila Pathak

8.1 INTRODUCTION

Traditional medicine serves as the most cost-effective and readily available form of treatment within the primary health care framework of economically disadvantaged communities. The local population has an extensive history of utilizing traditional plants for medicinal purposes, with evidence suggesting that plant therapy was used for a very long time—possibly as early as 4000–5000 B.C. Natural herbal measures as medications were first utilized by the Chinese. In India, plants were employed for medicinal purposes, as indicated in the Rig-Veda, a text believed to have been written between 1600–3500 B.C. Subsequently, ancient physicians in India systematically studied and empirically documented the properties and therapeutic applications of medicinal plants, forming the foundational knowledge of ancient medical science in the country. Throughout the world, indigenous medical systems rely heavily on medicinal plants, and ethnobotany offers a wealth of resources for the study and creation of natural drugs (Foster, 2021). The World Health Organisation (WHO) has created standards, guidelines, and strategies for botanical medicines in recognition of the importance of traditional medicine. It places a strong emphasis on using tried-and-true agro-industrial technologies to grow, prepare, and manufacture medicinal plants and herbal remedies.

One of the biggest and most varied families of angiosperms, orchids are well-known for both their exquisite flowers and therapeutic properties. As early as 200 B.C., the Chinese were the first to cultivate, describe, and use orchids (Singh and Duggal, 2009).

Cypripedium, commonly referred to as slipper orchids, stands out as one of the most esteemed members of the orchid family, boasting not only its captivating floral elegance but also a rich history of applications in traditional medicine (Rajeshbhai and Ingalhalli, 2022). In North America, using different species of *Cypripedium* has been a long-standing tradition. Both native people and immigrant communities have been using these orchids as sedative and antispasmodic qualities (Wilson, 2007). A notable species, *C. parvifolium*, held a significant role in traditional medicine, particularly in addressing issues related to insomnia and nervous tension. Historical records, such as Rafinesque's account from 1828, detail the meticulous collection of rhizomes during the fall or early spring, subsequent drying, and conversion into a therapeutic powder administered in a diluted form.

Samuel Thomson introduced the nerve powder, known as American Valerian, a concoction that featured three lady's slipper species (*C. acaule*, *C. parviflorum* var. *pubescens*, and C. *reginae*). This blend was applied broadly to address nerve-related afflictions and hysterical symptoms (Thomson, 1835). The integration of Slipper orchids into medicinal practices extends beyond North America, reaching Asia, where various parts of *Cypripedium* species, including stems, rhizomes, and flowers, are recognized for their analgesic, diaphoretic, sedative, and diuretic properties. These orchids have been employed in treating an array of ailments, such as rheumatic pain, dysentery, headaches, fractures, gastritis, leucorrhea, gonorrhea, and mental disorders (Cribb, 1997).

DOI: 10.1201/b23375-8

Despite the historical significance and diverse applications of *Cypripedium* in traditional medicine, comprehensive information on the medicinal values of these orchids remains limited and scattered, particularly about specific regions and communities everywhere in the world. This chapter's main goal is to compile and summarize the research findings that are currently available on various *Cypripedium* species, encompassing ethnopharmacology, phytochemistry, and conservation measures. By doing so, a comprehensive understanding of the therapeutic potential and cultural significance of slipper orchids can be achieved, contributing to the broader knowledge of these fascinating plants.

8.2 GEOGRAPHICAL DISTRIBUTION AND ITS DESCRIPTION

Cypripedium, commonly known as slipper orchids, is one of the most venerated in the orchid family. The genus has roughly 59 species and two varieties worldwide (Guo et al., 2012; Unruh et al., 2018) and is distributed in Alaska, Burma, Bhutan, Canada, China, Guatemala, Honduras, India, Japan, Mexico, Nepal, North Korea, Pakistan, Russia Federation, South Korea, Taiwan, and the United States. Cyprus and pedium are Latin words that refer to the slipper of Cyprus. According to Greek mythology, Aphrodite, the most beautiful of the Greek goddesses, lives on the island of Cyprus. *Cypripedium* species are herbaceous, small to large, and their inflorescences are racemose or almost spicate with very showy flowers (1–12 in number). Modifications of sepals and petals, presence of staminodes, columns with two anthers, and one-way trap flower (fixed and unidirectional route to pollinators) are amongst the features that make Cypripedioideae orchids special relative to all other orchids (Li et al., 2008). The fruit are linear ellipsoid capsules which produce hundreds of seeds. For the germination of orchid seeds in their natural habitat, the requirement of fungal associates is mandatory as they lack endosperm. The captivating charisma of these lady's slipper orchids distinguishes them as an ornamental plant not only by colour but also by the exclusivity of their specialized floral morphology, structure, and physiology.

8.3 THERAPEUTIC POTENTIAL OF *CYPRIPEDIUM* IN TRADITIONAL MEDICINE

Around the world, traditional medical systems have used various species of *Cypripedium* to treat a variety of illnesses, including dysentery, rheumatism, gastritis, fits, gonorrhea, headaches, leucorrhea, fractures, and mental disorders. In 1922, Harvey Wickes Felter documented that Cypripedium was identified as an optimal tranquilizer suitable for addressing states of nervous excitability or irritability, contingent upon the presence of atony. Felter (1923) highlighted its efficacy in dispelling gloom, instigating a serene and cheerful mental state, and promoting mental tranquility, ultimately fostering restful sleep. The product "Specific Medicine Cypripedium" was distributed by Lloyd Brothers, Pharmacists, Inc., based in Cincinnati, Ohio, who served as the principal supplier of prescription drugs exclusively prescribed by Eclectic physicians (Figure 8.1). Lady's slipper gained recognition as a primary drug in the United States Pharmacopeia (USP) as early as 1870 and was further integrated into the National Formulary (1916–1961). Several European and British Pharmacopeia also listed *Cypripedium pubescens* and *Cypripedium parviflorum* as nerve medicines (Teoh, 2019). *Cypripedium calceolus* is known as "Nerve Root" because it is used as a nerve tonic. It's interesting to note that the plant's "stinking & loathesome smell," which is identical to a goat's, gave rise to the moniker "Goat Stones" (Duggal, 1971). *Cypripedium himalaicum* is commonly used in the Sowa-Rigpa system of traditional medicine (Tsering et al., 2017), and *Cypripedium acaule* Aiton and *C. parviflorum* Salisb are used in North American folk medicine (De, 2023).

The root of *C. parvifolium* was found in extensive use in North America among both native and immigrant communities due to its sedative and antispasmodic qualities. It was commonly employed to alleviate concerns such as insomnia and nervous tension. The dried and powdered form of the *Cypripedium* rhizomes was administrated as a teaspoon of powder (Rafinesque and Alexander, 1828). Despite some aqueous whole-root preparations of *Cypripedium, they* were usually crafted

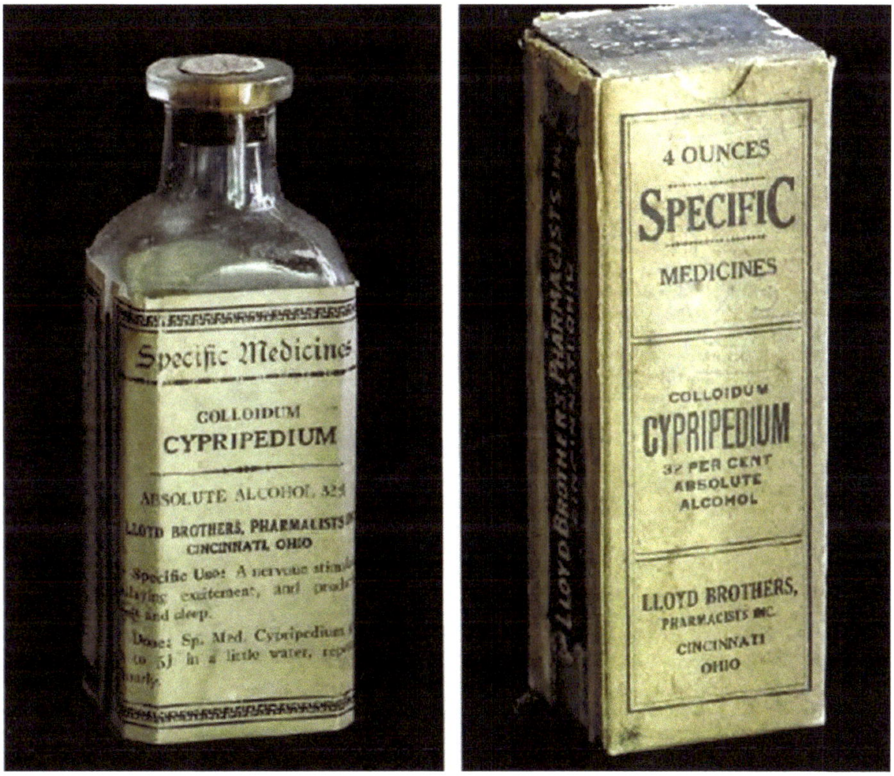

FIGURE 8.1 *Cypripedium* product manufactured by Lloyd Brothers, Pharmacist, Inc. (Cincinnati, Ohio) (Foster, 2021).

into tinctures due to non-water-soluble active principles. Tribes like the Cherokees, Iroquois, Menominee, MicMac, and Penobscot utilized *Cypripedium* for various disorders, including anxiety, convulsions, delirium, emotional tension, fever, headache, insomnia, irritable bowel syndrome, neuralgia, palpitations, tremors, as well as for easing menstrual and childbirth pain. Preparations of *Cypripedium* are widely recognized by herbalists for their sedative, tonic, hypnotic, anodyne, diaphoretic, nervine, antispasmodic, and stimulant qualities (Duke, 2002) (Figure 8.2). The different medical applications for *Cypripedium* are listed in Table 8.1.

8.4 BIOACTIVE COMPOUNDS IN *CYPRIPEDIUM* AND THEIR PHARMACOLOGICAL ACTIVITY

In the past few years, progress in extraction techniques and analytical methodologies has made it easier to characterize and identify bioactive compounds from medicinal plants, such as orchids. Medicinal orchids are rich reservoirs of diverse bioactive compounds, offering therapeutic potential in areas such as antibacterial, antioxidant, anti-inflammatory, anticancer, etc. The extensive biodiversity found in medicinal orchids provides a diverse range of chemical components, including alkaloids, flavonoids, terpenoids, phenolics, and essential oils, all of which harbour potential health benefits. These scientific methodologies have deepened our comprehension of the chemical and biological composition of therapeutic plants, paving the way for novel opportunities in drug discovery and development.

A perusal of the literature revealed that phenolics, alkaloids, and large amounts of potassium oxalate have been reported in *Cypripedium* species (Cribb, 1997). Khory and Collegeous

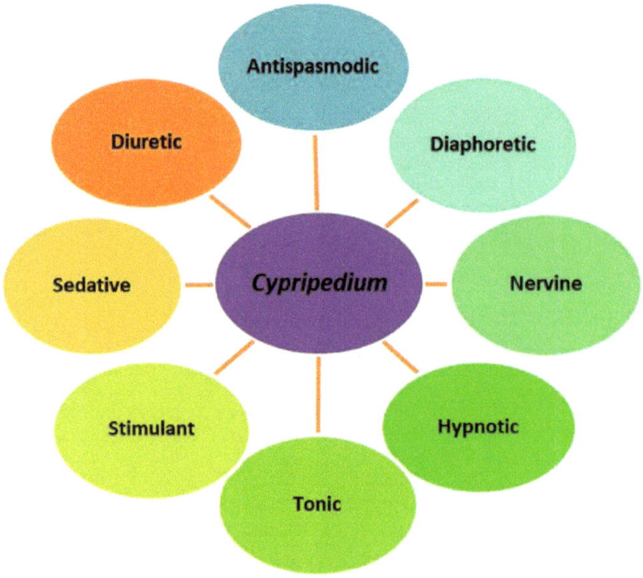

FIGURE 8.2 Therapeutic uses of *cypripedium* spp.

TABLE 8.1
Therapeutic Uses of a Few *Cyprepidium* Species

Name of the Species	Plant Part(s) Used	Therapeutic Uses	References
Cypripedium calceolus L.	Root, rhizome	Used as a sedative, to encourage slumber, and to lessen discomfort. For the treatment of headaches and jangling nerves, a tea (root) is made.	Singh and Somadey, 2005; Hossain, 2011
Cypripedium cordigerum D. Don	Root, Leaves	In Nepal, roots are utilized as a tonic and leaves as a vegetable.	Pant and Raskoti, 2013; Manandhar, 2002; Kumar and Singh, 2023
Cypripedium elegans Rchb. f	Root	Used to treat hysteria, spasms, fits, insanity, epilepsy, and rheumatism. Also used as an antispasmodic, sedative, diaphoretic, nervine, stimulant, and tonic.	Singh and Somadey, 2005; Hossain, 2011; Das, 2004; Baral and Kurmi, 2006; Jalal et al., 2010
Cypripedium debile Reichb. f.	Whole plant	In Taiwan, the entire plant is utilized as a diuretic and to cure pain and swelling..	Ou et al., 2003
Cypripedium fasciolatum Franch.	Root, Stem	Both stem and root are used as a diuretic and to treat fractures, edema, and joint pain.	Wu, 1994
Cypripedium formosanum Hayata	Root, Stem	Roots and stems are used to treat malaria, arthritis, and snake bites.	Lin et al., 2003
Cypripedium himalaicum Rolfe	Whole plant	The entire plant is used to treat coughs, urinary stones, coronary and pulmonary diseases.	Baral and Kurmi, 2006; Pant and Raskoti, 2013; Kumar and Singh, 2023

(Continued)

TABLE 8.1 *(Continued)*
Therapeutic Uses of a Few *Cyprepidium* Species

Name of the Species	Plant Part(s) Used	Therapeutic Uses	References
Cypripedium macranthos Sw.	Whole Plant Rhizome	The entire plant is used to treat coughs, urinary stones, and coronary and pulmonary diseases. It's also used for skin diseases.	Ou et al., 2003 Shimura et al., 2007; Kumar and Singh, 2023
Cypripedium parviflorum Salisbury	Rhizome	Powdered rhizome preparations are used to treat insomnia, anxiety, fever, headache, palpitations, irritable bowel syndrome, menstrual cramps, and childbirth pain.	Kumar et al., 2005; Hossain, 2011
Cypripedium pubescens Willd	Root	Roots are used as antispasmodic, diaphoretic, hypnotic, narcotic, and tonic and to cure convalescence, lack of healthy sustenance, diabetes, dysentery, diarrhoea, paralysis, impotence, and joint torment.	Singh and Duggal, 2009; Hossain, 2011
Cypripedium tibeticum King ex Rolfe	Root	In Tibet, Roots (6–9 g) is used to treat rheumatism, gonorrhoea, and leucorrhoea.	Wu, 1994

reported that the alcoholic extract of the plant contains active constituents 1–4 phenanthrenequinone, also known as cypripedin (Khory, 1903) (Figure 8.3F). Besides that other phenanthrenequinone compounds such as cypripedin and cypripediquinone A were also reported in the *Cypripedium* spp. (Ju et al., 2011; Krohn et al., 2001). Cypripedin demonstrated a mitigating effect on the characteristic mesenchymal traits, such as migratory behaviour, in non-small cell lung cancer H460 cells. This effect was accompanied by a notable decrease in actin stress fibres and focal adhesion, along with a weakened capacity for anchorage-independent growth. The findings suggest a promising pharmacological impact of cypripedin on lung cancer metastasis (Treesuwan et al., 2018).

Cypritibetquinones A and B were extracted from *Cypripedium tibeticum* King ex Rolfe's ethyl acetate extract. Their structures, identified as 7-hydroxy-2-methoxy-1,4-phenanthraquinone and 7-hydroxy-2,10-dimethoxy-1,4-phenanthraquinone, were determined through thorough spectral analyses having antioxidant properties (Figure 8.3 A and B) (Sritularak et al., 2011). Annoquinone is obtained from *Cypripedium macranthum,* which has anti-inflammatory and antimicrobial properties (Figure 8.3C). Cypripedin and Cryptostylin are reported from *Cypripedium calceolus*, *Cypripedium macranthum* (De, 2023). From the seedlings of *Cypripedium macranthos* var. rebunense, two antifungal compounds, namely lusianthrin and chrysin, were extracted. These compounds exhibited a slightly stronger antifungal activity relatively specific to the non-pathogenic *Rhizoctonia* spp (Figure 8.3 D and E) (Shimura et al., 2007). The antimicrobial activity of *C. cordigerum* (from leaves, roots, and tubers) was evaluated in ethanolic, methanolic, chloroform, and aqueous extracts against *Pseudomonas aeruginosa* and *Staphylococcus aureus*. The findings showed that all extracts at a concentration of 100 mg/ml were active against *P. aeruginosa* and *S. aureus*, with inhibition ranging from 7 to 16 mm and 7 to 16 mm, respectively (Bharal et al., 2014).

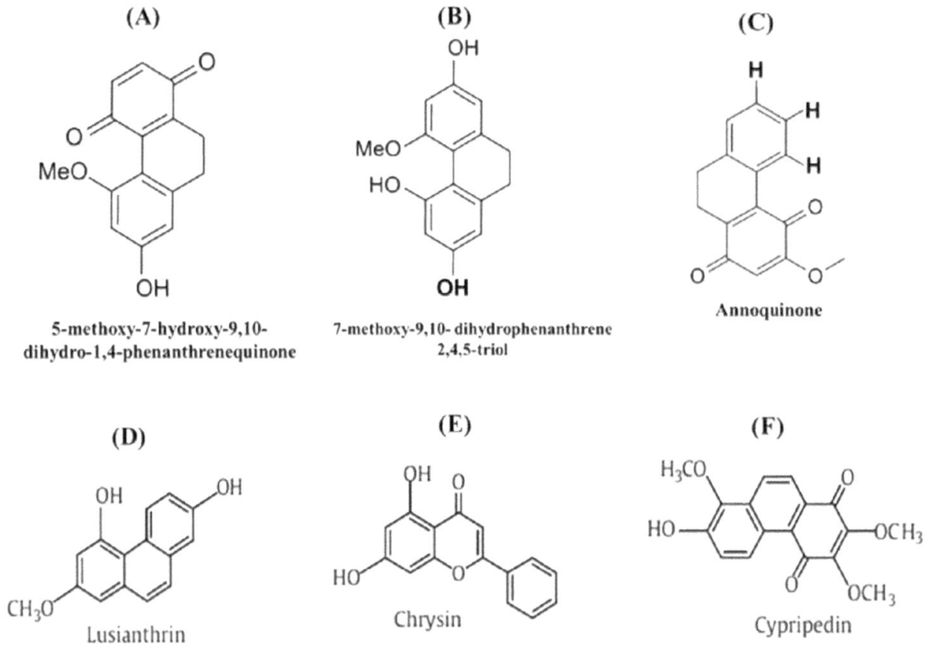

FIGURE 8.3 Bioactive Compounds reported in *Cypripedium* spp.

8.5 THREATS AND CONSERVATION STRATEGIES

The native populations of these species are experiencing a decline, attributed to a range of factors like habitat destruction, urbanization, competition from non-native species, mining, livestock grazing, climate change, and various disturbances. One of the main factors causing these species populations to decline is the complex pollination process, which is marked by low rates of fruit set and uneven germination beneath natural conditions. Furthermore, these species are environmentally sensitive and require specific habitat conditions for germination and growth, such as the presence of mycorrhizal fungi, nutrient availability, and sunlight, as described by Kull in 1999. International Union for Conservation of Nature (IUCN) reported that 56.5% of orchids are classified as vulnerable, endangered, or critically endangered. Cypripedioideae is the only orchid subfamily at particularly high risk, with 90% of assessed species falling into one of the threat categories (Fay and Rankou, 2016). Moreover, the complete Orchidaceae family is included in Appendix II of CITES. Preserving this potential medicinal resource is crucial for ensuring its long-term availability for human use.

Most terrestrial orchid seeds germinate in natural settings only after being infected by species-specific mycorrhizal fungi, as described by Stewart et al. (2008). A decline in favourable soil conditions, as well as a reduction or loss of fungi in their habitat, can impede orchid plant regeneration and have a negative impact on in situ conservation efforts (Arditti and Ghani, 1999; Batty et al., 2001). As a result, it is essential to develop ex-situ conservation protocols for orchids as part of a comprehensive strategy that links germplasm storage with plant reintroduction into their native environment. Most of the terrestrial orchid seeds only germinate in nature following infection by mycorrhizal fungi, an interaction that can be species-specific (Stewart et al., 2008). A decrease in favourable soil conditions and reduction or loss of fungi from the habitat may become major constraints for orchid plant regeneration and negatively affect *in situ* conservation efforts (Arditti and Ghani, 1999; Batty et al., 2001). Therefore, it is necessary to develop ex-situ conservation protocols for orchids as part of an integrated strategy linking the storage of germplasm with the reintroduction of plants into their natural habitat. The technique of vitro tissue culture stands as a practical alternative for the swift mass multiplication of orchids and finds extensive application in the preservation of endangered species. Knudson developed a method for the asymbiotic germination of orchid seeds, i.e., without fungal association (Stewart et al., 2008). The morphogenesis of *Cyperidium acaule* was studied *in vitro* (Leroux et al., 1995). The subfamily Cypripedioideae comprises four genera (*Cypripedium, Paphiopedilum, Phragmipedium*, and *Selenipedium*), all of which are recognized as challenging to culture *in vitro*, with *Cypripedium* being the most challenging (Arditti in 2008). Chu and Mudge (1994) have found in *Cypripdeium calaeolus* that the practice of prechilling was advantageous and liquid medium supported the highest levels of germination. As reported by de Pauw and Remphrey in 1993, seeds collected at eight weeks after pollination (WAP) showed a significant decline in germination percentage across three *Cypripedium* species. In contrast, Jiang et al. (2017) identified the peak germination percentage for *C. lentiginosum* to occur in seeds collected at 90–105 days after pollination (DAP) during the early globular to globular stage of embryo development. According to recent research, (Ghimire et al., 2020), seeds collected around 9–10 WAP or 65–75 DAP could prove beneficial for *in vitro* seed germination of *C. japonicum*.

The present modern biotechnological methods, including the *in vitro* micropropagation technique, hold tremendous potential for the production of high-quality plant-based medicine, large-scale multiplication of disease-free plants, faster cloning, and the conservation of desired genotypes in a very short period. Current advanced biotechnological approaches, such as the *in vitro* micropropagation technique, possess significant potential for generating high-quality plant-based medications, large-scale plant multiplication free of disease, rapid cloning, and preserving desired genotypes within a short timeframe. However, less information is available on *in vitro* conservation of *Cyprepidium* species; therefore, further deep studies are required to conserve the species.

8.6 CONCLUSION

Exploring the ethnobotanical aspects of *Cypripedium* sheds light on the remarkable array of therapeutic properties inherent in these plants, encompassing anodyne, antispasmodic, diaphoretic, tonic, nervine, hypnotic, sedative, and stimulant attributes. Despite the rich traditional knowledge surrounding their medicinal use, there are notable gaps in pharmacological studies that warrant comprehensive exploration. Bridging these gaps is imperative to fully harness the therapeutic potential embedded in orchids of the *Cypripedium* genus. Recent studies in modern biology have unveiled promising antimicrobial and antifungal properties within the chemical constituents of these plants. This discovery not only aligns with traditional applications but also opens avenues for innovative research in developing novel therapeutics.

Current research endeavours hold the potential to identify new biomolecules within *Cypripedium* species, thereby providing scientific validation for the efficacy of conventional treatments. This synergy between traditional wisdom and contemporary scientific exploration offers a promising avenue for the discovery and development of potent therapeutic agents. As we explore the pharmacological complexities of *Cypripedium* orchids, the potential exists to discover innovative compounds that could revolutionize the field of medicine, providing fresh and effective treatments for a spectrum of health conditions.

REFERENCES

Arditti, J. 2008. *Micropropagation of Orchids*, 2nd ed. Oxford: Blackwell Publishing Ltd.

Arditti, J. & Ghani, A. 1999. Numerical and physical properties of orchid seeds and their biological implications. *New Phytologist* 145:367.

Baral, S. R. & Kurmi, P. P. 2006. *A Compendium of Medicinal Plants in Nepal*. Kathmandu: Mass Printing Press.

Batty, A., Dixon, K., Brundrett, M. & Sivasithamparam, K. 2001. Long-term storage of mycorrhizal fungi and seed as a tool for the conservation of endangered Western Australian terrestrial orchids. *Australian Journal of Botany* 49:619–628 (https://doi.org/10.1071/BT01029).

Bharal, A., Kashyap, M., Sohpal, V. K. & Sembi, J. 2014. Evaluation of antimicrobial properties of terrestrial orchids against certain human pathogens. *International Journal of Bioassays* 3:3036–3039.
Chu, C. & Mudge, K. 1994. Effects of prechilling and liquid suspension suspension culture on seed germination of the yellow lady's slipper orchid. *Lindleyana* 9:153–159.
Cribb, P. 1997. *Genus Cypripedium*. Portland, OR: Timber Press. International Plant Names Index.
Das, S. 2004. Indian orchids in indigenous medicine system. In: Brirto, S. J. (Ed.), *Orchids Diversity and Conservation*. Tiruchirappalli: The Rapinat Herbarium, St John's College.
De, L. 2023. Pharmaceutical properties of orchids LC De. *The Pharma Innovation* 12:1454–1460.
De Pauw, M. & Remphrey, W. 1993. *In vitro* germination of three *Cypripedium* species in relation to time of seed collection, media, and cold treatment. *Canadian Journal of Botany* 71:879–885.
Duggal, S. C. 1971. Orchids in human affairs (a review). *Quarterly Journal of Crude Drug Research* 11:1727–1734 (https://doi.org/10.3109/13880207109066247).
Duke, J. A. 2002. *Handbook of Medicinal Herbs*, 2nd ed. Boca Raton: CRC Press (https://doi.org/10.1201/9781420040463).
Fay, M. & Rankou, H. 2016. Slipper orchids on the IUCN Red List. In: Stuart, S. (Ed.), *2015 Annual Report to the Environment Agency – Abu Dhabi. Framework Support for Implementing the Strategic Plan of the IUCN Species Survival Commission*. Gland: IUCN Species Survival Commission.
Felter, H. 1923. The eclectic materia medica. *Pharmacology and Therupeitlics* 12:188–189 (https://doi.org/10.1002/jps.3080120233).
Foster, S. 2021. Ladys slipper: Once a commercial conundrum, now a conservation success story. *HerbalGram* 131:40–51.
Ghimire, B., Son, S., Kim, J. H. & Jeong, M. J. 2020. Gametophyte and embryonic ontogeny: Understanding the reproductive calendar of *Cypripedium japonicum* Thunb. (Cypripedoideae, Orchidaceae), a lady's slipper orchid endemic to East Asia. *BMC Plant Biology* 20:426 (https://doi.org/10.1186/s12870-020-02589-9).
Guo, Y.-Y., Luo, Y.-B., Liu, Z.-J. & Wang, X.-Q. 2012. Evolution and biogeography of the slipper orchids: Eocene Vicariance of the Conduplicate Genera in the old and new world tropics. *PLoS One* 7:e38788 (https://doi.org/10.1371/journal.pone.0038788).
Hossain, M. 2011. Therapeutic orchids: Traditional uses and recent advances—An overview. *Fitoterapia* 82:102–140 (https://doi.org/10.1016/j.fitote.2010.09.007).
Jalal, J. S., Kumar, P., Tewari, L. & Pangtey, Y. P. S. 2010. Orchids: Uses in traditional medicine. In: *Medicinal Plants of Himalaya: Potential and Prospects*. India: Regional Research Institute of Himalayan Flora, Tarikhet, Uttarakhand, pp. 128–136.
Jiang, H., Chen, M. C. & Lee, Y. I. 2017. In vitro germination and low-temperature seed storage of *Cypripedium lentiginosum* PJ Cribb & SC Chen, a rare and endangered lady's slipper orchid. *Scientia Horticulturae* 225:471–479 (https://doi.org/10.1016/j.scienta.2017.07.040).
Ju, J., Yang, J., Li, J. & Xiao, P. 2011. Cypripediquinone A, a new phenanthraquinone from *Cypripedium macranthum* (Orchidaceae). *Chinese Chemical Letters* 11:37–38.
Khory, N. 1903. Materia medica of India and their therapeutics. *The Indian Medical Gazette* 38:476–477.
Krohn, K., Loock, U., Paavilainen, K., Hausen, B., Schmalle, H. & Kiesele, H. 2001. Synthesis and electrochemistry of annoquinone-A, cypripedin methyl ether, denbinobin and related 1,4-phenanthrenequinones. *Arkivoc* 1:88–130.
Kull, T. 1999. *Cypripedium calceolus* L. *Journal of Ecology* 87:913–924 (https://doi.org/10.1046/j.1365-2745.1999.00407.x).
Kumar, A. & Singh, L. 2023. Ethnomedicinal uses of orchids and its species under threat: A review. *Research Journal of Pharmacy and Technology* 16:4469–4478 (https://doi.org/10.52711/0974-360X.2023.00729).
Kumar, P., Pandey, A., Rawat, G. & Jalal, J. 2005. Diversity and conservation of Orchids in state of Jharkhand. In: *Plant Taxonomy: Advances and Relevance*. New Delhi: CBS Publ. and Distr, pp. 345–353.
Leroux, G., Barabe, D. & Vieth, J. 1995. Comparative morphogenesis of *Cypripedium acaule* (Orchidaceae) protocorms cultivated in vitro with or without sugar. *Canadian Journal of Botany* 73:1391–1406.
Li, P., Luo, Y. B., Deng, Y. X. & Kou, Y. 2008. Pollination of the lady's slipper Cypripedium henryi Rolfe (Orchidaceae). *Botanical Journal of the Linnean Society* 156:491–499 (https://doi.org/10.1111/j.1095-8339.2007.00767.x).
Lin, I., Chang, Y., Chen, I. & Hsieh, W. 2003. *The Catalogue of Medicinal Plant Resources in Taiwan*. Kuala Lumpur: China Press, p. 153.
Manandhar, N. 2002. *Book Review: Plants and People of Nepal*. Portland: Timber Press.
Ou, J., Hsieh, W., Lin, I. & Chang, Y. 2003. *The Catalogue of Medicinal Plant Resources in Taiwan*. Taipei Oudhia: Department of Health, Executive Yuan.

Pant, B. & Raskoti, B. 2013. *Medicinal Orchids of Nepal.* Kathmandu: Himalayan Map House Pvt. Ltd.

Rafinesque, C. S. & Alexander, A. 1828. *Medical Flora, or, Manual of the Medical Botany of the United States of North America.* Philadelphia: Atkinson & Alexander (https://doi.org/10.5962/bhl.title.524).

Rajeshbhai, V. & Ingalhalli, R. 2022. A review article on: An overview of medicinal importance of orchids. *Acta Scientific Medical Sciences* 6:28–36 (https://doi.org/10.31080/ASMS.2022.06.1306).

Shimura, H., Matsuura, M., Takada, N. & Koda, Y. 2007. An antifungal compound involved in symbiotic germination of *Cypripedium macranthos* var. rebunense (Orchidaceae). *Phytochemistry* 68:1442–1447 (https://doi.org/10.1016/j.phytochem.2007.03.006).

Singh, A. & Duggal, S. 2009. Medicinal orchids: An overview. *Ethnobotanical Leaflets* 2009:3.

Singh, M. & Somadey. 2005. *Indian Medicinal Plants*, 1st ed. New Delhi: Satish Serial Publishing House.

Sritularak, B., Anuwat, M. & Likhitwitayawuid, K. 2011. A new phenanthrenequinone from *Dendrobium draconis. Journal of Asian Natural Products Research* 13:251–255 (https://doi.org/10.1080/10286020.2010.546354).

Stewart, S., Kauth, P., Dutra Elliott, D., Johnson, T., Kane, M. & Vendrame, W. 2008. Techniques and applications of in vitro orchid seed germination. In: *Floriculture, Ornamental and Plant Biotechnology: Advances and Topical Issues.* Global Science Books, UK, pp. 375–391.

Teoh, E. 2019. *Orchids as Aphrodisiac, Medicine or Food.* Switzerland: Springer (https://doi.org/10.1007/978-3-030-18255-7).

Thomson, S. 1835. *New Guide to Health, or, Botanic Family Physician: Containing a Complete System of Practice, Upon a Plan Entirely New; With a Description of the Vegetables Made Use of, and Directions for Preparing and Administering Them to Cure Disease; to Which is Prefixed a Narrative of the Life and Medical Discoveries of the Author—Digital Collections—National Library of Medicine.* Boston: J.Q. Adams, Printer.

Treesuwan, S., Sritularak, B., Chanvorachote, P. & Pongrakhananon, V. 2018. Cypripedin diminishes an epithelial-to-mesenchymal transition in non-small cell lung cancer cells through suppression of Akt/GSK-3β signalling. *Scientific Reports* 8:8009 (https://doi.org/10.1038/s41598-018-25657-5).

Tsering, J., Tam, N., Tag, D., Gogoi, B. & Apang, O. 2017. Medicinal orchids of Arunachal Pradesh: A review. *Bulletin of Arunachal Forest Research* 32(1&2):1–16.

Unruh, S., Mckain, M., Lee, Y.-I., Yukawa, T., McCormick, M., Shefferson, R., Smithson, A., Leebens-Mack, J. & Pires, J. 2018. Phylotranscriptomic analysis and genome evolution of the Cypripedioideae (Orchidaceae). *American Journal of Botany* 105:631–640 (https://doi.org/10.1002/ajb2.1047).

Wilson, W. 2007. *Medicinal Plant Fact Sheet: Cypripedium: Lady's Slipper Orchids.* Virginia: Arlington.

Wu, X. 1994. *A Concise Edition of Medicinal Plants in China.* Guangdong: Guangdong Higher Education Publication House (in Chinese).

9 Medicinal and Therapeutic Uses of *Dendrobium* Orchids

T. Yatung, Kalkame Ch. Momin, V. Bhargav, S. Pooja Murthy, and T. Usha Bharathi

9.1 INTRODUCTION

Dendrobium stands as the second-largest genus within the Orchidaceae family, encompassing a range of 900–2,000 species. India harbors 116 *Dendrobium* species, with 70 identified in North Eastern India, situated in the Indo-Burma hotspot, the world's second-largest biodiverssity hotspot. The remaining species are distributed in the Western Ghats of India (Deori et al., 2019). Notably, over 80 species have been documented in China (Wang, 2021). Among all the species, the orchids within the *Dendrobium* genus are recognized for containing the highest levels of phytochemicals. They exhibit diverse bioactivities, encompassing antidiabetic, antitumor, antiviral, antioxidant, anti-aging, anti-inflammatory, gastrointestinal protective, immunomodulatory, cardio-protective, and anti-osteoporosis properties (Chen et al., 2021).

Dendrobium species have long been utilized in traditional Chinese medicine to nurture the physique and to address various skin ailments. Notably, several species, including *Dendrobium officinale, D. chrysanthum, D. fimbriatum, D. loddigesii, D. phalaenopsis,* and *D. nobile* are widely utilized in the preparation of herbal medicines in China (Wang, 2021). The aphrodisiac properties of *Dendrobium* species are acknowledged, with *D. nobile, D. densiflorum, D. jenkinsii,* and *D. aduncum* being incorporated into the formulation of the Chinese tonic 'Shi Hu' in traditional medicine (Ninawe and Swapna, 2017). In the Indian state of Meghalaya, Khasi and Garo Tribes use the juice of the crushed leaves of *Dendrobium moschatum* Wall., as ear drops to cure ear pain (Rao, 1981). In the southern part of Gujarat state in India, the local tribes use the crushed leaves of *D. microbulbon* as ethno-medicinal herb for the stomach ache (Sharma et al., 2007). The Mishing tribe of Arunachal Pradesh use the leaf paste of *D. amoenum* to treat skin diseases as well as to hasten the healing of bone fractures and reduce swelling (Shankar et al., 2012).

Dendrobium orchids have spiritual and cultural significance in India and specifically *Dendrobium nobile, D. hookerianum,* and *D. gibsonii* are cultivated in the monasteries of Arunachal Pradesh to signify the sanctity and purity (Hegde et al., 2017). Many *Dendrobium* species have been in the food chain, besides its medicinal, cultural, and aesthetic importance. The leaves, young shoots, and pseudobulbs of *Dendrobium* are used in the preparation of various cuisines as well as cattle feed to improve cattle health and higher milk production. The dried leaves of *D. salacceme* are cooked with rice by the Tamang community of Nepal to add delicate and exotic flavor to the food. Likewise, pseudobulbs of *Dendrobium speciosum* and *D. microbulbon* are used as food, and flowers of *D. longicornu* are used as pickles.

In Chinese, 'Tie pishihu' is the common name for *D. officinale* stems, which are highly priced and utilized in traditional medicine (Ye et al., 2017), health drinks, nutraceuticals, beverages (Zhu et al., 2018; Meng et al., 2019). Besides, it is frequently used in culinary preparations owing to the richness of pectin and glucomannans, which account for its water-absorbing, thickening, and emulsifying properties (Cakova et al., 2017; Devaraj et al., 2019). In countries like Phillipines, Indonesia, and New Guinea, *Dendrobium* species are extensively valued for the preparation of items

 DOI: 10.1201/b23375-9

like baskets from the stems. Similarly, in India, *Dendrobiums* are utilized in craft work; for example, the dried golden stems of *D. formosum* are used by the tribes of Andaman and Nicobar islands to prepare waist bands, baskets, decorative items, etc. (Awasthi, 1991). Flowers of *D. densiflorum* are popularly used by tribes of Manipur to decorate weapons.

Breeders have cultivated and hybridized wild orchids, resulting in the creation of numerous *Dendrobium* hybrids featuring a variety of colors, shapes, and forms. These hybrids are highly sought after in both domestic and international markets, specifically for the purpose of cut flowers and potted plants (Chen and Tsi (2000)). The vibrant-colored flowers of *Dendrobium* also make them well-suited for use as dry flowers and have great potential in the floriculture industry.

9.1.1 Botany

The botanist Olof Swartz described the genus *Dendrobium* in 1799, representing a huge collection of orchids that predominantly reside as epiphytes on tree branches. The name *Dendrobium* is derived from the Greek terms 'dendron' (meaning tree) and 'bios' (meaning life), indicating that this genus primarily thrives in tree habitats, classifying it as an epiphytic group. *Dendrobium*, an epiphytic orchid, is the dominant genera among commercially important orchid genera and finds a prime position in the production of commercial cut flowers. *Dendrobium* is the second-largest genus within the orchid family, showcasing a diverse array of shapes, colors, and morphological characteristics. There is a distinguished degree of heterogeneity among the intrageneric taxa within this genus. *Dendrobium* is a sympodial epiphytic orchid with over 1,400 species. They exhibit close relationships with genera such as *Bulbophyllum*, *Flickingeria*, and *Eria*. However, it distinguishes itself by featuring free lateral sepals and four pollinia organized into two sets of two, lacking appendages like caudicles or viscidia (Fig 9.1). The flowers include a distinctive 'chin' formed by the lip, column foot, and lateral sepals. This unique combination of mentum and necked pollinia sets the *Dendrobium* genus apart. *Dendrobium* plants have peculiar long canes, otherwise called pseudobulbs, with soft, fleshy leaves all along the entire length of the plant, while some may have short or swollen pseudobulbs that terminate in two coriaceous leaves. There are variations in the type of pseudobulbs, which may be classified as woody cane, cylindric, clavate fleshy, and bulbous round (Fig 9.4). Leaves are usually thick, with the size ranging from 2.50 to 40 cm, and are either deciduous or evergreen (De et al., 2015b) (Fig 9.2 & 9.3). The inflorescence in *Dendrobiums* is terminal or sub-terminal in position with single to many blooms per cane and of varying colours, size and length. While in some species of *Dendrobium*, flowers are borne solitary, small/minute in size that emerges from the axil of the leaves. The arrangement of the flowers are either in pairs or in three on short peduncles along the length of the pseudobulbs and occasionally surrounded by caudecous leaves.

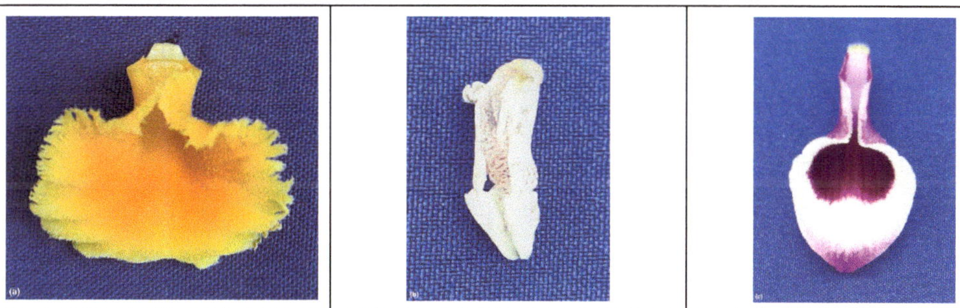

Orbicular (*D. chrysotoxum*)	Oblanceolate (*D. macraei*)	Ovate (*D. nobile*)

FIGURE 9.1 Variation in lip shape of *Dendrobiums*.

Lanceolate (*D. aphyllum*)	Ovate (*D. densiflorum*)	Elliptic (*D. nobile*)

FIGURE 9.2 Variation in leaf shape of *Dendrobiums*.

Acute (*D. devonianum*)	Acuminate (*D. chrysanthum*)	Obtuse (*D. macraei*)	Retuse (*D. eriifolium*)

FIGURE 9.3 Variation in leaf apex of *Dendrobiums*.

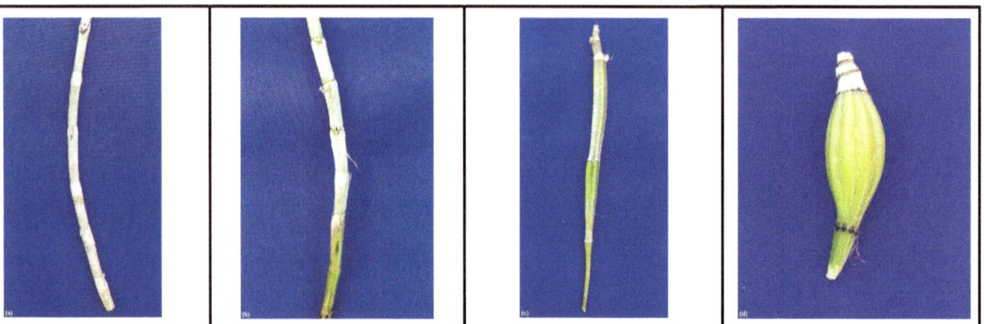

Cane (woody) (*D. aphyllum*)	Cane (cylindric fleshy) (*D. chrysanthum*)	Cane (clavate fleshy) (*D. densiflorum*)	Bulbous (round) (*D. chrysotoxum*)

FIGURE 9.4 Variation in nature of shoot/cane of *Dendrobiums*.

FIGURE 9.5 *Dendrobium* orchids with medicinal uses.

9.2 SPECIES

Next to the genus *Bulbophyllim*, *Dendrobium* constitutes the second largest group in Orchidaceae with over 1,536 species worldwide and is distributed mostly in the tropical and subtropical regions of South East Asia like India, China, Japan, Malaya, Philippines, New Guinea, and also in Australia, Pacific Islands, and New Zealand. There exists a remarkable diversity in terms of their habit and habitat, ranging from high altitude to even dry desert conditions. The large, robust, and attractive flowers of the genus *Dendrobium*, are commercially cultivated for ornamental purposes and are becoming popular amongst flower lovers owing to their exquisite flower form and longevity (Fig 9.5). Besides their ornamental value, certain species, like *Dendrobium nobile, D. chrysotoxum, D. moschatum*, etc., have been used by traditional healers for their medicinal value in addressing certain health problems of the tribal community worldwide. Further, species like *Dendrobium aggregatum, D. chrysotoxum, D. formosum, D. nobile, D. primulinum*, and *D. williamsoni*, have been used in crop improvement programs which have resulted in the development of many present-day hybrids.

9.2.1 Origin and Distribution

Dendrobium species have wide adaptability, ranging from the dry desert areas to the snow-clad mountain regions. *Dendrobium* is native to South East Asia and can be found in their natural habitat from North East India to the islands of New Guinea and Polynesia (Table 9.1). The genus *Dendrobium* comprises of mostly epiphytic, occasionally lithophytic species of nearly 1,100 species reported worldwide (Wood, 2006). In India, *Dendrobium* is the second-largest orchid group after *Bulbophyllum*, comprising 103 species distributed across the Western Himalayas, Eastern Himalayas, Western Peninsular, and Andaman and Nicobar Islands (Singh et al., 2001). Out of these, as many as 82 species of *Dendrobium* have been reported from the North Eastern region alone. Arunachal Pradesh has the maximum diversity with 49 species of *Dendrobium*, followed by Manipur (48), Asam (46), Meghalaya (42), Sikkim (41), Mizoram (40), Nagaland (30), and Tripura (5) (Lokho, 2013).

Besides the North Eastern region of India, a great number of *Dendrobium* species are also found in the hot and humid regions of West Bengal and the hilly regions of Bihar and Orissa in India. A considerable number of species are found in the Kumaon Hills, comprising Nainital, Almora, Garhwal, and Tehri districts. Central and Southern India is also home to a large number of *Dendrobium* species, particularly in Konkan, Kanara, Malabar, and the Nilgiris Hills. *Dendrobiums* are also found in abundance in the Andaman and Nicobar Islands in the Eastern Bay of Bengal and the Nilgiris Hills.

TABLE 9.1
Distribution of *Dendrobium* Species Worldwide

S. No.	Country	Important *Dendrobium* Species	References
1.	China	*D. acinaciforme, D. aduncum, D. aphyllum, D. aurantiacum, D. bellatulum, D. brymerianum, D. chameleon, D. changjiangense, D. christyanum, D. chrysanthum, D. chrysotoxum, D. compactum, D. dixanthum, D. guangxiense, D. hancockii, D. hainanense, D. loddigesii, D. moniliforme, D. porphyrochilum, D. strongylanthum, D. officinale, D. nobile*	Pan et al. (2022), Yang et al. (2020), Zhong and Yang (2021)
2.	Indonesia	*D. acuminatissimum, D. crumenatum, D. linearifolium, D. macrophylum, D. heterocarpum, D. secundum, D. aloifolium, D. truncatum, D. salacsense, D. spathilingue, D. stuartii, D. rugosum, Dendrobium lamellatum, D. anosmum, D. bracteosum, D. purpureum, D. mirbelianum, D. inflatum, D. lancifolium, D. hamatipes*	Isa et al. (2018), Hartati et al. (2022), Clements (2003)

(Continued)

TABLE 9.1 *(Continued)*
Distribution of *Dendrobium* Species Worldwide

S. No.	Country	Important *Dendrobium* Species	References
3.	Myanmar	*Dendrobium naungmungense, D. ciliatilabellum, D. vexabile, D. clavator, D. draconis, D. dixanthum, D. parcum, D. delacourii, D. aphrodite, D. calothyrsos, D. hirtulum, D hkinhumense, D. koyamae, D. laterale, D. leucochlorum, D. luteolum, D. marmoratum, D. pedilochilum, D. praetermissum, D. sarmentosum, D. rhodocentrum*	Liu et al. (2018), Htwe (2019), Aung et al. (2020)
4.	Philippines	*Dendrobium orbilobulatum, D. planum, D. rhombeum, D. setigerum, D. tiongii, D. boosii, D. microphyton, D. auriculatum, D. guereroi, D. sanderae, D. orbilobulatum, D. philippinense, D. balzerianum, D. crumenatum, D. polytrichum, D. linguella, D. marivelense, D. quisumbingii, D. ramosii, D. salaccense, D. sanguinolentum, D. schuetzei, D. secundum, D. sibuyanense, D. papilio, D. yeagari, D. alsiosum, D. cumingii*	Aljohn and Kier (2020), Dizon et al. (2018), Agoo (2007), Fischer et al. (2007), Clements (2003)
5.	New Guinea	*Dendrobium canaliculatum, D. carronii, D. d'albertisii, D. devosianum, D. discolor, D. ionoglossum, D. johannis, D. nindii, D. semifuscum, D. sylvanum, D. trilamellatum, D. lasienthera, D. phlox, D. lawesii, D. aemulans, D. chloroleucum, D. spectabile, D. falce, D. capituliflorum, D. smillae, D. bracteosum, D. rigidum, D. inequale, D. luteocillium, D. williamsianum, D. insigne, D. acerosum, D. forbesii, D. retroflexum, D. johnsoniae, D. uncinatum, D. dryatum, D. oreocharis, D. fulgidum, D. pseudo-frigidum, D. anosmum, D. confusum, D. malbrownii, D. teretifolium, D., roseipes, D. cancroides, D. glomeratum, D. atroviolaceum, D. chrysoglossum, D. bigibbum, D. cincinnatum, D. dichaeoides, D. trachyrhizum, D. coeloglossum, D. rhodostichtum, D. warianum, D. macrophyllum, D. ornatum*	Arobaya et al. (2020), Millar (1978), Jones (2018)
6.	Laos	*Dendrobium aduncum, D. aloifoloium, D. anosmum, D. aphyllum, D. chryseum, D. callipes, D. chrysotoxum, D. cumulatum, D. curviflorum, D. crepidatum, D. compactum, D. deltatum, D. densiflorum, D. exile, D. farmeri, D. devonianum, D. friedricksianum, D. hercoglossum, D. gratiosissimum, D. indivisum, D. langbianens, D. loddigesii, D. loddigesii, D. pachyglossum, D. oligophyllum, D. parciflorum, D. ruckeri, D. salaccense, D. schildhaueri*	Schuiteman et al. (2008)
7.	Vietnam	*D. fimbriatum, D. findlayanum, D. moschatum, D. anosmum, D. parishii, D. chrysanthum, D. primulinum, D. hancockii, D. tortile, D. amabile, D. thyrsiflorum, D. farmeri, D. daoense, D. chrysotoxum, D. callipes, D. wattii, D. lindleyi, D. nobile, D. draconis, D. brymerianum, D. bellatuium, D. cariniferum, D. ochraceum, D. unicum, D. virginium, D. wardianum, D. williamsonii, D. cruentum, D. harveyanum*	Duong et al. (2018), Averyanov and Averyanova (2003)
8.	Sri Lanka	*Dendrobium aphyllum, D. crumenatum, Dendrobium nutantiflorum, D. haemoglossum, D. heterocarpum, D. maccarthiae, D. panduratum*	Sandamali et al. (2020), Fernando and Ormerod (2008)
9.	Malaysia	*Dendrobium crumenatum, D. aloifolium, D. anosmum, D. bancana, D. bifarium, D. concinnum, D. convexum, D. excavatum, D. farmeri, D. grande, D. leonis, D. metrium, D. pachyphyllum, D. plicatile, D. sanguinolentum, D. secundum, D. singaporense, D. sinuatum, D. subulatum, D. villosulum, D. xantholeucum, D. candidum, D. nobile, angustifolium, D. foxii, D. hughii, D. longipes, D. uniflorum*	Moudi and Go (2015), Go et al. (2011), Isa et al. (2018)

(Continued)

TABLE 9.1 ***(Continued)***
Distribution of *Dendrobium* Species Worldwide

S. No.	Country	Important *Dendrobium* Species	References
10.	Thailand	*Dendrobium aloifolium, D. formosum, D. hercoglossum, D. palpebrae, D. lindleyi, D. signatum, D. fridericksianu, D. chrysanthum, D. lituiflorum, D. anosmum, D. finlayanum, D. albosanguineum, D. parishii, D. primulinum, D. unicum, D. capillipes, D. dixanthum, D. ellipsophyllum, D. cruentum, D. formosum, D. infundibulum, D. virgineum, D. christyanum, D. scabrilingue, D. crumenatum, D. compactum, D. acerosum*	Clements (2003); Fischer et al. (2007), Peyachoknagul et al. (2014)
11.	Nepal	*Dendrobium amoenum, D. aphyllum D. anceps, D. bicameratum, D. candidum, D. chrysanthum, D. darjeelingense, D. denneanum, D. densiflorum, D. denudans, D. eriiflorum, D. fimbriatum, D. moniliforme, D. formosum, D. moniliforme, D. monticola, D. gibsonii, D. hookerianum, D. longicornu, D. moschatum, D. nobile, D. ochreatum, D. peguanum, D. porphyrochilum, D. transparens, D. polyanthum, D. pulchellum*	Rokaya et al. (2013)

9.3 PHYTOCHEMISTRY

Dendrobium orchids are known to contain approximately 131 bioactive compounds, which include bibenzyls, flavonoids, lignans, phenanthrenes, phenylpropanoids, alkaloids, polysaccharides, and several other compounds. These compounds are noticed for their multiple pharmacological properties such as antioxidant, antimicrobial, anti-inflammatory, anticancer, anti-psoriasis, anti-aging, immunomodulatory, hypoglycemic, and tyrosinase inhibitory activities (Pan et al., 2014; Wang, 2021).

1H-Indole-3-carbaldehyde and 3-hydroxy-4,5-dimethoxypropiophenone are the first compounds isolated from the orchids under the *Dendrobium* genus. The bioactive compound phenanthrenes exhibit antifibrotic, anticancer, anti-inflammatory, and antibacterial activities. Stilbenes is one of the bibenzyl compounds, are reported to possess antioxidant, anti-migratory, antimutagenic, anti-inflammatory, and retinal neo angiogenesis inhibitory effects as reported by Hwang et al. (2010), Lin et al. (2013), Kim et al. (2015), and Chen et al. (2015). Additionally, syringic acid, a naturally produced O-methylated trihydroxybenzoic acid monomer, exhibits antidiabetic and antioxidant activities (Yang et al., 2006, 2007; Zhang et al., 2008; Hwang et al., 2010; Wei et al., 2012; Chen et al., 2013; Lin et al., 2013; Kim et al., 2015; Vinayagam et al., 2016; Zhou et al., 2016).

Of the different isolated compounds, phenanthrenes are the most prominent ones (Yang et al., 2007; Hwang et al., 2010; Lin et al., 2013; Chen et al., 2013; Klongkumnuankarn et al., 2015; Kim et al., 2015; Zhou et al., 2016). A detailed study on *D. chrysanthum* showed that the plant is rich in phenanthrenes like apocynin, chrysotoxin, confusarin, 5,7-dimethoxyphenanthrene-2,6-diol, dendrochrysanene, ephemeranthol-A, moscatilin, nudol, orchinol. Besides, phenols (3-hydroxy-4,5-dimethoxypropiophenone, moscatin, syringaldehyde, trans-3,4,5-trimethoxylcinnamyl alcohol), alkaloids (1-(6-hydroxy-[3]pyridyl)-propan-1-one, carbofuran), benzofurans (loliolide), monoterpenes (6R,7aR)-7a-hydroxy-3,6-dimethyl-5,6,7,7a-tetrahydrobenzofuran-2(4H)-one) and lignans (3,3′-bis(3,4-dihydro-4-hydroxy-6-methoxy-2H-1-benzopyran), (+)-syringaresinol), cinnacassin F, 1-(6-hydroxy- [3] pyridyl)-propan-1-one, (+)—(1S,5R)-verbenon-10-oic acid, were also present in the plants (Yang et al., 2006). Carbofuran which is used as a broad spectrum, systemic insecticide as well as nematicide and ((6R,7aR)-7a-hydroxy-3,6-dimethyl-5,6,7,7a-tetrahydrobenzofuran-2(4H)-have been isolated from *D. chrysanthum* (Zhang et al., 2022). Denbinobin B isolated from *D. sinense* has been shown to have moderate antibacterial activity against *Staphylococcus aureus* (Chen et al., 2013). The stems of *D. denneanum* are used for isolation of benzoquinones, bibenzyls, β-sitosterol, cinnamic acid derivatives, coumarins, daucosterol and phenanthrenes (Yang et al., 2006; Lin et al.,

2013). Dry stems of *D. aurantiacum* var. *denneanum* comprises of coumarin, chrysotobibenzyl, dendroflorin, dengibsin, gigantol, kaempferol, moscatilin, naringenin, taraxerol (Yang et al., 2006). *D. gratiosissimum* is found to have naturally occurring bibenzyl derivatives such as batatasin III, dengraols A, dengraols B. dihydroresveratrol, erianin, gigantol, moscatilin, tristin, and 3,4'-dihydroxy-5-methoxybibenzyl that show cytotoxic activities against tumor cells (Zhang et al., 2008). Stem of *D. nobile* comprises of about 17 phenanthrenes such as 2-hydroxy-4,7-dimethoxy-9,10-dihydrophenanthrene, 2,2'-dihydroxy-3,3',4,4',7,7'-hexamethoxy-9,9',10,10'-tetrahydro-1,1'-biphenanthrene, 2,3,5-trihydroxy-4,9-dimethoxyphenanthrene, 2,8-dihydroxy-3,4,7-trimethoxyphenanthrene, 2,8-dihydroxy-3,4,7-trimethoxy-9,10-dihydrophenanthrene, 3-hydroxy-2,4,7-trimethoxy-9,10-dihydr ophenanthrene, and 3,4,8-trimethoxyphenanthrene-2,5-diol which has antifibrotic effect on the tissues (Yang et al., 2007 and Hwang et al., 2010).

9.4 THERAPEUTIC USES

The existence of bioactive compounds is precisely associated with the medicinal properties of plants (Duraipandiyan et al., 2006). Orchids within the *Dendrobium* species contain bioactive compounds, including alkaloids, anthraquinones, bibenzyls, coumarins, fluoro-derivatives, phenanthrenes, and sesquiterpenes (Table 9.2). These bioactive compounds have been reported to exhibit antioxidant, antifibrotic, anti-inflammatory, anti-platelet aggregation, and immunomodulatory activities (Lin et al., 2013). The isolation of bioactive compounds from plants has paved the way for their application in modern medicine to treat various ailments. Traditional Chinese medicine recognizes *'Dendrobii Herba'* as containing four phenolic compounds—syringic acid, vanillic acid, ferulic acid, and 4-hydroxybenzoic acid. The presence of nine phenanthrenes is attributed to the anti-inflammatory activity of the plant against cytokines that cause inflammation (Yoo et al., 2017). Bioactives like cyclotetracosane, caryophyllene, morin, α-humulene oleic acid, n-hexadecanoic acid, etc. in the leaf extracts account for the antioxidant and anticancer properties of *D. aphyllum* (Rahamtulla et al., 2023).

D. denneanum extracts are reported to cause suppression of lipopolysaccharide (LPS)-induced expression of inducible NO synthase (iNOS) which have been linked to the presence of active ingredients like and 5-methoxy-2,4,7,9S-tetrahydroxy-9,10-dihydrophenanthrene, 2,5-dihydroxy-4-methoxy-phenanthrene 2-O β-D glucopyranoside. Besides, they also inhibit kappa B α (I κ B α) and phosphorylation of JNK, p38, mitogen-activated protein kinase (MAPK), showing their anticancer anti-inflammatory effects by hindering nuclear factor κ B (NF κ B) synthesis and MAPKs pathways (Lin et al., 2013). In *D. chrysanthum*, the compound dendrochrysanene exhibits anti-inflammatory activities by causing suppression in the mRNA levels of iNOS, IL8, IL10, and TNF α in murine peritoneal macrophages (Yang et al., 2006).

A study by Xiong et al. (2013), confirmed the neuroprotective activity, cytotoxicity against tumor cells, and glutamate-induced toxicity in PC12 cells by *D. aurantiacum* var. denneanum, owing to the presence of a plethora of active components like 8,4' oxyneolignane 7-O-glucoside, (−)-syringaresinol-4,4'-bis-O-β-D-glucopyranoside, (−)-(7S,8R,7'E)-4-hydroxy-3,3',5,5' tetramethoxy-8,4'-oxyneolign-7'-ene-7,9,9'-triol 7,9'-bis-O-β-D-glucopyranoside, syringaresinol-4'-O-D-monoglucopyranoside, yringin, 1-[4-(β-D-glucopyranosyloxy)-3,5-dimethoxyphenyl]-1 propanone, shashenoside I, and vicenin-2. The presence of dendroflorin, denchrysan, gigantol, flavanthrinin, moscatilin, lusianthridin, nobilone, and tristin in *D. brymerianum* is known to exhibit antimigratory and cytotoxic effects, reflecting their use as potential anti-metastatic drugs (Klongkumnuankarn et al., 2015).

Several *Dendrobium* species contain compounds with cytotoxic action against cancer cells, such as Erianin from *D. chrysotoxum*, Dengrasol A and B from *D. gratiosissimum* (Zhang et al., 2008; Li et al., 2001). *D. officinale* showcases anticancer, anti-inflammatory, antitumor, and antioxidant properties due to the presence of flavonoids like Apigenin-6-C-1α-L-arabinoside-8-C-1β-D-xyloside, Apigenin-6,8-di-C-1α-L-arabinoside, Apigenin-6-C-(2''-O-1β-D-glucopyranoside)-1α-L-arabinoside, Apigenin-6-C-1β-D-xyloside-8-C-1β-D-arabinoside, Apigenin-6-C-1β-D-xyloside-8-C-1β-D-glucoside, Isoschaftoside, Isoviolanthin, Naringenin, Rutin, Schaftoside, Syringaresinol, Vicenin-2, Vicenin 3,

TABLE 9.2
Medicinal Uses of *Dendrobium* Orchids

Sl.No.	Name of Orchid	Distribution	Parts Used	Uses	Bioactive Compound	References
1.	*Dendrobium amoenum* Wall. ex. Lindl.	Himalayas	Pseudobulb	Fresh paste is administered to burned skin and dislocated bones. Antibacterial and antioxidative properties.	Isoamoenylin	Singh et al. (2012).
2.	*Dendrobium aphyllum* (Roxb.) C.E.C. Fisch/*Dendrobium cucullatum*	Northeastern India	Leaves	It is used to treat leucorrhoea, rheumatism, wounds, earaches, epilepsy, paralysis, depression, abdominal pain, and inflammation of the eyes. It is applied to kids whose cranial structures are malformed. It has antioxidant properties.	n-hexadecanoic acid, oleic acid, flavone, morin, caryophyllene, cyclotetracosane, α-humulene, Aphyllone B, bibenzyls, and phenanthrenes	Akhter et al. (2017).
3.	*Dendrobium aurantiacum* Var. *denneanum* (Kerr) Z.H. Tsi	Southern China and South East Asia	Stems and leaves	Cytotoxic effect, anticancerous	Moscatalin	Sliwinski et al. (2022).
4.	*Dendrobium bellatulum* Rolfe	Northeastern India and South East Asia.	Whole plant	Used to treat mouth dryness, heat damage after illness, poor vision, gagging, stomach pain, and bright vision.	2-hydroxy-4-methoxy-3,6-dimethylbenzoic acid, aloifol I, batatasin III, dendrosinen B, dendrosinen D, p-hydroxycinnamic acid, neoolivil	Shang (2019).
5.	*Dendrobium brymerianum* Rchb.f.	South East Asia	Whole plant	Cytotoxic	Moscatalin, gigantol, lusianthridin and dendroflorin	Sliwinski et al. (2022).
6.	*Dendrobium candidum*	Eastern Asia, Southeast Asia, and Australia	Stem	Warming the stomach, soothing the throat, regulating blood pressure, boosting immunity, improving vision, strengthening the kidneys, and extending life are some of its benefits. Biological properties include improving immunity, preventing liver damage, inhibiting cancer, lowering blood sugar, reducing inflammation, and preventing diabetes.	Polysaccharide, dendrobine, and flavonoids (flavone)	Li and Li (2017), Li et al. (2017).

7.	*Dendrobium catenatum/ Dendrobium officinale* Kimura et. Migo	All over the world (Japan, United States, China, and Australia)	Stem	Support the production of fluids, tone the stomach, nourish the heart, and treat fever. Biological actions such as decreasing blood sugar and blood lipids, improving immunity, and preventing cancer. It also strengthens the heart, balances the five viscera, gets rid of arthralgia, eases weariness, makes the stomach thicker, makes the body lighter, and increases longevity. It has biological functions like antitumor, cardio-protective, anti-aging, antidiabetes, antioxidant, gastrointestinal protective, immunomodulatory, and anti-osteoporosis effects.	Glucomannan, β-glucosidases, Bibenzyls, ephemeranthol A, erianthridin, orchinol, phenylpropanoids, coumarin, flavonoids (apigenin, vitexin, quercetin, and kaempferol), densiflorol A. 3,4'-Dihydroxy-5-methoxybibenzyl, Dendrocandin F,G,H,I,J,K (Bibenzyls)	Qi et al. (2022), Chen et al. (2021), Wang (2021).
8.	*Dendrobium chrysanthum* Wall. ex. Lindl.	China, Himalayas, and Southeast Asia	Leaves	Leaves are used as an antipyretic and to treat mild skin conditions, while the stem is used as a tonic to strengthen the immune system, lower fever, and benefit the eyes. Anti-inflammatory.	Dendrochrysanene	De et al. (2015), Choudhary et al. (2023), Wang (2021).
9.	*Dendrobium chryseum* Rolfe.	Himalayas	stem	Antioxidant	Quercetin	Wang (2021).
10.	*Dendrobium crumenatum* Sw	Malaysia	Leaves	Boils and pimples are treated with a leaf poultice.	Cyclododecane, 1-decene, 2,4-bis (1,1-dimethylethyl) phenol, 3,5-dimethylbenzaldehyde and octadecanoic acid	Wang (2021), Wongjuk and Thirabunyanon (2020)
11.	*Dendrobium chrysotoxum* Lindl.	Himalayas, Northeastern India, China, and Southeast Asia	Stems, flowers, leaves	Extracts of stem and flowers are used as tonic. It possesses antipyretic, anti-inflammatory, antibacterial, anti-psoriasis, anti-cataract hypoglycemic, and anti-angiogenic properties.	Scoparone, Erianin	Tsering et al. (2017), Singh et al. (2012), Ramesh et al. (2019), Wang (2021).

(Continued)

TABLE 9.2 *(Continued)*
Medicinal Uses of *Dendrobium* Orchids

Sl.No.	Name of Orchid	Distribution	Parts Used	Uses	Bioactive Compound	References
12.	*Dendrobium crepidatum* Lindl. and Paxton	Northeastern India	Pseudobulb, stem	Bones that are shattered or dislocated are treated with pseudobulb paste. Rheumatism and arthritis are treated using stem leaves. Anti-inflammatory, anti-carcinogenic, and antioxidant.	Dendrocrepidine A, B, C, D, E, and F, homocrepidine A and B, crepidine, Isocrepidamine	Tsering et al. (2017), Hu et al. (2018), Paudel et al. (2019), Wang (2021).
13.	*Dendrobium denneanum* Kerr (syn. *Dendrobium aurantiacum* var. *denneanum* [Kerr] Z.H.Tsi)	China	Stems, leaves	Impetigo is treated using stems. External application of dry leaves for impetigo treatment. Anti-inflammatory	Moscatin, Phenanthrenes	Wang (2021)
14.	*Dendrobium densiflorum* Lindl.	Northeastern India	Pseudobulbs, leaves	Pseudobulb relieves thirst, treats vomiting, and lowers phlegm and fever. Rejuvenates the digestive system's weakened capacity brought on by a decrease in appetite and digestive fluids. Pseudobulbs' pulp is used to treat boils, zits, and other skin problems. Leaf paste is applied to sprains, inflammations, and fractured bones. Biological properties include antimicrobial and anti-inflammatory properties. In vitro ant-platelet aggregation activity is reported.	Dendroflorin, phenanthrenes, gigantol, scopoletin, moscatilin, scoparone, densiflorol A and B, moscatin, confusarin, dendrodensiflorol, cypripedin	Fan et al. (2001), Li et al. (2023), Kovács et al. (2008), Singh (2022).
15.	*Dendrobium denudans* D. Don	Northeastern India	Stem	Used to treat tonsillitis, colds, coughs, and nasal obstructions. Relieves bodily pains and elevated fever. Additionally, it is used as a tonic for kids and the elderly. Possessing antioxidant qualities.	Flavonoids (Rutin), Phenols, alkaloids	Tsering et al. (2017), Singh et al. (2015), Bhattacharya et al. (2020).
16.	*Dendrobium devonianum* Paxton	Northeastern India, Bhutan, Myanmar, Thailand, and China	Stem	An immune booster	Flavonoids(quercetin, kaempferol), Phenolic acids, dendrophenenes A-B, retusiusine B, and amoenylin	Tsering et al. (2017), Li et al. (2023).

17.	*Dendrobium draconis*	Southeast Asia	Aerial parts	Antioxidant and anti-carcinogenic	Gigantol, hircinol	Wang (2021), Sliwinski et al. (2022).
18.	*Dendrobium ellipsophyllum*	Burma, Thailand, Laos, Vietnam, and Yunnan Province, China	Whole plant	Anti-carcinogenic and anti-metastatic activities	5,7-Dihydroxy-chromen-4-one, 4,5-Dihydroxy-2,3-dimethoxy-9,10 –dihydrophenanthrene, Moscatilin, -Homoeriodictyol, (2S)-Eriodictyol, Chrysoeriol, Phloretic acid, Luteolin	Tanagornmeatar et al. (2014).
19.	*Dendrobium falconeri* Hook.	Northeastern India	Stem	Consumed as a tonic and used to cure lung cancer. Has anti-herpetic properties.	Bis(bibenzyls) like dendrofalconerols A and B, Phenolic compounds include docosanoyl (E)-ferulate, tetracosyl (Z)- p -coumarate, tetracosyl (E)- p -coumarate, 2-(p -hydroxyphenyl)ethyl p -coumarate, p -hydroxybenzoic acid, and p -hydroxybenzaldehyde. 2-(p -Hydroxyphenyl)ethyl p -coumarate	Tsering et al. (2017), Sritularak and Likhitwitayawuid (2009)
20.	*Dendrobium findlayanum* Par. and Rchb.f.	Chinese Himalayas, Myanmar, and Thailand	Stems, whole plant	Anticancerous	Dendrobine, Benzyl derivatives	Liu et al. (2020)
21.	*Dendrobium fimbriatum* Hook.	Himalayas, China, and Vietnam	Whole plant, pseudobulbs (paste/ extract)	Whole plant is utilized for neurological debility and disturbed liver. Pseudobulb is considered to have stimulant, aphrodisiac, and demulcent properties. Pseudobulb is used to treat fever, nervous debility, and disturbed liver. Leaf decoction can be used as a tonic to encourage bodily fluids. The shattered bones are set using leaf paste.	Fimbriatone, moscatilin, gigantol, defuscin, chrysotoxine, 2,5-phenanthrenediol, 9–10-dihydro-7- methoxy, fimbriadimerbibenzyls	Tsering et al. (2017), Ramesh et al. (2019).

(Continued)

TABLE 9.2 *(Continued)*
Medicinal Uses of *Dendrobium* Orchids

Sl.No.	Name of Orchid	Distribution	Parts Used	Uses	Bioactive Compound	References
22.	*Dendrobium fugax* Rchb.f.	Southeast Asia, India, and Sri Lanka	Tubers, whole plant	Powder is used as a general debility stimulant and tonic. Used to treat sleeplessness and as an aphrodisiac.	Salutaridine, thebaine	Tsering et al. (2017), Singh (2022).
23.	*Dendrobium heterocarpum* Wall. ex. Lindl.	Southeast Asia, Nepal, and India	Pseudobulbs	Broken and misaligned bones are treated using paste. Biological characteristics include anti-inflammatory and anti-obesity effects.	Amoenylin, methyl 3-(4-hydroxyphenyl) propionate, 3,4-dihydroxy-5,4'-dimethoxybibenzyl, 3-Hydroxy-4',5-dimethoxybibenzyl dendrocandin B, dendrofalconerol A, and syringaresinol	Warinhomhoun et al. (2022), Wang (2021) and Longchar and Deb (2021).
24.	*Dendrobium herbaceum* Lindl.	India	Roots	After burning fresh roots, 10 g of the ash is combined with 10 ml of mustard oil and applied to the afflicted area two to three times a day for many days or until the symptoms go away.	Phenanthrenes, alkaloids, flavonoids, phenylpropanoids, and lignans, loddigesiinol A, (S)-5- methoxy-2,4,7,9-tetrahydroxy-9,10-dihydrophenanthrene	Wang (2021)
25.	*Dendrobium huoshanense* C.Z.Tang et S.J.Cheng	China	Stem	Stems are often used to produce bodily fluid, ease stomach discomfort, remove heat, and replenish yin, or body in Chinese. Immunoregulation, antioxidant, anti-cataract, anti-glycation, anti-aging, antitumor, anti-rheumatoid arthritis, anti-atherosclerosis, anti-inflammation, hypoglycemia action, and liver protective activities are examples of biological functions.	Polysaccharides, flavonoids, bibenzyls, phenols, sesquiterpenoids	Pan et al. (2014), Gao et al. (2022).
26.	*Dendrobium infundibulum* (Lindl.) Kuntze	Northern Indo- China	Whole plant	Cytotoxic effect	Ephemeranthol A	Sliwinski et al. (2022).

27.	*Dendrobium jenkinsii* Wall. ex. Lindl.	Northeastern India, Bhutan, Myanmar, and China	Stems	Chinese medicine Shih-hu is prepared using both fresh and dried stems. Used to cure boils, acne, cholera, and nerves. Used to treat conditions such as gastritis, thrush, dehydration during fever, and dry eyes.	Flavonoids	De et al. (2015), Bhattacharya et al. (2020), Tsering et al. (2017).
28.	*Dendrobium loddigesii*	China, Indo-China	Stems	Anti-aging (collagen production), anti-inflammatory, antioxidant, anti-carcinogenic, and antimutagenic.	Batatasin III, 3,3',5-Trihydroxybibenzyl, 4.4-Dihydroxy-3,5-dimethoxybibenzyl, Gigantol, Tristin, Crepidatin, Moscatilin, Moscatilin diacetato, Loddigesiinol C, D, Trigonopol B	Singh et al. (2012), Li et al. (2019), Wang (2021).
29.	*Dendrobium longicornu* Lindl.	Southern China, Himalayas, and Northern Indo-China	Whole plant, roots, protocorm	Juice from plants is used to treat fever. Boiling roots are used to alleviate coughing cattle.	Tetracosanoic acid, 9-Hexadecyn-1-ol, 3-Heptadecanol, Pentafluoropropionatetrans-2-dodecen-1-ol, 6,10-Dimethyl-4-undecanol, α-Cadinol	Tsering et al. (2017), Paudel et al. (2020).
30.	*Dendrobium macraei* Lindl.	Himalayas	Tubers	As a general stimulant and demulcent, paste is used to prevent snake bites. Eczema is treated by applying paste to the skin's afflicted area. Adrenaline-containing remedy used for fever, throat problems, sinusitis, and asthma. Tonic for overall weakness.	A and β jibantic acid and an alkaloid jebantine	Singh and Duggal (2009), Pant (2013), Wang (2021).
31.	*Dendrobium moniliforme* (L.) Sw.	Himalaya and temperate Eastern Asia	Pseudobulb	Antioxidant, antipyretic action	Dendrobine, Phenylpropanoids (isomers of ferulate)	Singh et al. (2012), Wang (2021)

(Continued)

TABLE 9.2 *(Continued)*
Medicinal Uses of *Dendrobium* Orchids

Sl.No.	Name of Orchid	Distribution	Parts Used	Uses	Bioactive Compound	References
32.	*Dendrobium nobile* Lindl.	Himalayas, Southeast Asia, and China	Whole plant (powder)	Tonic, vigor, aphrodisiac, stomachic, analgesic, and anti-aging properties are all associated with stems. Stem tonic is used to lessen menstruation discomfort, nocturnal sweats, thirst, tongue dryness, and salivation. Stem powder is used for dyspepsia, lumbago, fever, parched and thirsty mouth, anorexia, and pulmonary tuberculosis. Utilized whole in neurological diseases. Seeds are put on newly cut wounds to promote rapid healing. Biological effects include those that are antioxidant, anti-inflammatory.	Copacamphane, confusarin, chrysotoxine, crepidatin, cyclococamphane, Denbrobin A, Dendroflorin, Dendrocoumarin, coelonin, Noblin A, B, C, and D, pinoresinol, Flavanthrinin, dendramine, dendrophenol, gigantol, nobiline, picrotoxane, moscatilin, Dendronobiloside A, C, D, E, F, Fimbriol B	Singh and Duggal (2009), Rahman and Huda (2021), Singh et al. (2012), Tsering et al. (2017), Wang (2021), Hwang et al. (2010), Zhang et al. (2008).
33.	*Dendrobium normale*	Southeast Asia, Australia, Pacific Islands	Whole plant	The whole plant is utilized for its aphrodisiac and tonic qualities.	Flavonoids, alkaloids, glycosides, phytosterols, triterpinoids	Jalal et al. (2008), Rao and Babu (2017).
34.	*Dendrobium ovatum*	Western ghats of India	Stem	Juice obtained by hand crushing of stems is used to treat constipation and stomachache. Biological activities like Hepatoprotective. The plant is used as an emollient.	Flavonoids, Phenols, alkaloids, triterpenoids, glycosides, steroids, and carbohydrates	De et al. (2015), Ganapaty et al. (2013), Wang (2021).
35.	*Dendrobium officinale*	Southeast Asia	Whole plant	Juice of the whole plant is obtained by ethanol extraction used to treat cancer cells.	Polysaccharide	Haizhen et al. (2017)
36.	*Dendrobium palpebrae*	Bangladesh, Northeastern India, Indian subcontinent, tropical Asia	Whole plant	Antioxidant	Dendroflorin	Wang (2021).
37.	*Dendrobium secundum*	Indo-China, South East Asia, and Andaman and Nicobar Islands.	Stem	Antioxidant	Moscatilin, syringaresinol	Wang (2021).

38.	*Dendrobium signatum* Rchb.F.	Southeast Asia	Leaves	Anti-carcinogenic	Dendrocandin B, I and dendrofalconerol, 3,4,-dihydroxy-3,4,-methoxybibenzyl	Sliwinski et al. (2022).
39.	*Dendrobium thyrsiflorum*	Himalayas and mountains of Northern Indo-China	Whole plant	Anti-carcinogenic	2.7- Dihydroxy-4-methoxyphenanthrene, 2,7-Dihydroxy-4-methoxy-9-fluorenone, 3,7-Dihydroxy-2,4-dimethoxyphenanthrene, 2,3,5-Trihydroxy-4-methoxy-9,10-dihydrophenanthrene, Dengibsin, Denchrysan B	Sliwinski et al. (2022).
40.	*Dendrobium tortile* Lindl.	Southeast Asia, Bangladesh, Assam state in India, and Andaman Islands.	Whole plant	The herb has tonic and aphrodisiac properties. It has antioxidant and anti-inflammatory properties. Decreases insulin resistance and functions as an α-glucosidase inhibitor.	Eriodictyol, trans-Tetracosylferulate, cis-Docosylferulate, 3,4-Dihydroxy-3,4′-dimethoxybibenzy, Dendrofalconerol A	Rahman and Huda (2021), Limpanit et al. (2016).
41.	*Dendrobium transparens* Wall. ex. Lindl.	North-east India, Nepal, and Myanmar	Pseudobulb	Broken and misplaced bones are treated using pseudobulb paste. Anti-carcinogenic biological action.	10-octadecenoic acid methyl ester, 15-methyl-hexadecanoic acid methyl ester, 17-octadecynoic acid, Hexadecane	Tsering et al. (2017), Joshi et al. (2020).
42.	*Dendrobium venustum*	Southeast Asia	Whole plant	Antimalarial, anti-carcinogenic and anti-herpetic properties	Flavanthrinin, Gigantol, Densiflorol B, Lusianthridin, Phoyunnanin E, Phoyunnanin C, E, and Dihydroartemisinin	Sukphan et al. (2014), Sliwinski et al. (2022).
43.	*Dendrobium wardianum* Warner	Southeast Asia	Stems	Cytotoxic effect	Dendrocandin V, phenanthrenes, moscatalin, loddigesiinols A	Sliwinski et al. (2022).
44.	*Dendrobium williamsonii*	Northeastern India and Thailand	Whole plant	Antioxidant, anti-carcinogenic	3,3′-Dihydroxy-4,5-dimethoxybibenzyl, moscailin, Apigenin	Wang (2021)

Violanthin, Vitexin-2”-O-1*β*-D-glucopyranoside, lignans, and phenolic acids such as Ferulic acid, p-Hydroxycinnamic acid, Syringic acid, Vanillic acid, and Dihydroconiferyldihydro-p-coumarate (Ye et al., 2017). The presence of Isoviolanthin in *D. officinale* leaves has been observed to reduce the migratory and invasive abilities of Hepatocellular carcinoma (HCC) cells treated with TGF-β1. This discovery suggests potential applications for treating advanced-stage metastatic HCC, as reported by Xing et al. (2018).

It has been reported that syringic acid in *'Dendrobii Herba'* causes downregulation of aldose reductase mRNA, thereby inhibiting its activity. Syringic acid is also known to stimulate β-cell regeneration in the pancreas, rectify anomalies in glycoprotein, and facilitates condensation reactions and oxidation polymerization, hence playing a prominent role in the treatment of diabetes (Wei et al., 2012; Vinayagam et al., 2016). In *D. aphyllum* stems, the presence of compounds like mosscatin, tricetin 3’,4’,5’-trimethyl ether 7-O-β-glucopyranoside, and moscatilin causes nitric acid inhibition, prevents cytotoxicity and inflammation (Yang et al., 2015). *D. huoshanense* stems are rich in polysaccharides (glucose, galactose, xylose, traces of galacturonic acid) that show cataractogenesis, hypoglycemic, and anti-glycation activities and have been traditionally used to treat throat infections, and chronic superficial (Pan et al., 2013, 2014). *D.* officinale is a well known medicinal orchids that exhibits hypoglycemic activity, anti-inflammatory activity against tumor necrosis factor (TNF)-α and interleukin (IL)-1β, cardio-protective activity against oxidative stress diabetes-induced cardiomyopathy (Pan et al., 2014; Xu et al., 2014; Hou et al., 2016; Zhang et al., 2016).

In the Chinese traditional system of medicine, two medicines, namely ‘Tiepi shihu’ and ‘Shihu,’ are derived from the stems of *Dendrobium* orchids (Jin et al., 2015). The latter is a formulation of 30 species of *Dendrobium*, including *D. chrysotoxum, D. fimbriatum, D. nobile*, and other related species, while the former specifically refers to preparation from *D. officinale* (Cakova et al., 2017). *D. chrysotoxum*, a component of ‘Shihu,’ is well-recognized for its role in enhancing retinal angiogenesis and anti-hyperglycemic activities. The plant is famously known to heal chronic vascular complications that might lead to blindness by inhibiting the NFκB signaling pathway (Pan et al., 2014; Gong et al., 2014; Yu et al., 2015).

9.5 TRADITIONAL USES

The diverse *Dendrobium* species have been integral to traditional medicinal practices, including traditional Chinese Medicine, Indian Ayurveda, and Korean and Japanese medicine, across various cultures throughout history (Pant et al., 2022). Orchids of *Dendrobium* species have been in use as a medicinal plant for centuries by practitioners of the traditional Chinese medicine system and documented as a “superior grade” herbal medicine in the ancient text *Shen Nong's Herbal Classic* (Xu et al., 2013a), of which *D. officinalis* is the most expensive species. The dried stem of D. officinalis, known in Chinese as Tie-Pi-Feng Dou, is traditionally used to treat chronic gastritis, hyperglycemia, hyperlipidemia and poor eyesight. It is valued for its immuno boosting effects and is also recognised for its anticancer and antiaging properties (Chen et al., 2013; Chinese Pharmacopoeia, 2015). Zi-Pi-Feng-Dou prepared from *D. devonianum* has similar morphology and efficacy as *D. officinalis* and is used as an adulterant owing to the high price of *D. officinalis* (Cheng et al., 2015).

The traditional Chinese and Korean medicine systems utilize *'Dendrobii Herba,'* obtained from the dried stems of *Dendrobium* species such as *D. loddigessi, D. nobile, D. chrysanthum, D. chrysotoxum, D. candidum/D. moniliforme, D. fimbriatum var. oculatum,* and *D. catenatum/D. officinale* (Lam et al., 2015; Gu et al., 2017). These herbs are employed as antipyretics and for the treatment of conditions like atrophic bone and tendon issues, atrophic gastritis, blurred vision, cardiovascular diseases, diabetes, dry throat, and gastrodynia, and for balancing ‘yin’ while promoting the production of body fluid (Takamiya et al., 2011; Xu et al., 2013b; Ye et al., 2017; Yoo et al., 2017). The stems of *Dendrobium aurantiacum* var. *denneanum* known as ‘Huangcao’ or ‘Shihu’ in Chinese, are traditionally used in the Chinese traditional medicine system as an anti-aging, antioxidant, antipyretic, immunomodulatory, and for improving eyesight (Xiong et al., 2013). The

Dendrobium plants are extensively used in Chinese traditional medicine as well as finding applications in clinical medicine, which is leading to the extinction of the plants of this genus from the wild, making them endangered in China and thus promoting the cultivation of *D. denneanum* for isolation of bioactive compounds to be used as medicine (Lin et al., 2013). *Dendrobium* species have been used to heal the wide range of ailments, one such includes the skin disorders. The skin diseases are often persistent and difficult to treat, may cause mortality though the mortality percentage is very low when compared to other diseases (Wagh and Jain, 2020). Many herbal plants have been used traditionally to treat the dermatological disorders world over, orchids species have also been used for the same. In India, China, Nepal, Australia, Liberia, Bangladesh, Indonesia, and Malaysia, 22 *Dendrobium* species are employed locally for the treatment of dermatological disorders, as noted in the New Inventory of Existing Cosmetic Ingredients in China. Specifically, six *Dendrobium* species, namely, *D. nobile, D. loddigesii, D. chrysanthum, D. bigibbum* var. superbum, *D. fimbriatum*, and *D. officinale* have been listed in the IECIC 2015 Final Version. *Dendrobium* finds its application in treatment of many dermatological disorders; for example, in Australia, the natives use the sap from the bulb or stem and pseudobulb of *D. affine* to directly apply to treat cuts, itchy skin, minor cuts, and sores (Smith, 1991; Tucci and Wilkens, 2016). In India, Bangladesh, and China, the application of paste or juice from the leaves and the whole plant of *D. aphyllum* is a common practice for treating burns, wounds, scalds, and various skin infections. Additionally, it finds usage in traditional medicine for ailments such as earaches, epilepsy, craniosynostosis, and paralysis (Zhang and Zhang, 1994; Li, 1995; Hossain, 2009; Akhter et al., 2017; Rahamtulla et al., 2020). Various species of *Dendrobium*, including *D. affine, D. alpestre, D. crumenatum, D. densiforum, D. discolor, D. fimbriatu*, and *D. monticola*, are applied for boils. *D. affine, D. fimbriatum* and *D. nobile* are used for cuts, while *D. affine, D. aphyllum, D. amoenum, D. nobile* and *D. polyanthum* are employed for burns. *D. macraei* and *D. polyanthum* find application in treating eczema, *D. denneanum* for impetigo, *D. purpureum* for nail infections and *D. densiforum, D. monticola, D. alpestre, D. crumenatum*, and *D. fimbriatum* for acne. Moreover, *D. polyanthum* and *D. aphyllum* are used for treating scalds, *D. macraei* and *D. macrostachyum* for skin allergies, *D. chrysanthum, D. affine*, and *D. canaliculatum* for sores, *D. discolor* for ringworm, *D. hancockii* for ulcers and *D. nobile, D. aphyllum, D. amoenum, D. fimbriatum*, and *D. chrysanthum* for wounds (Wang, 2021). *Dendrobium* species are also utilized in culinary practices to enhance immunity, prevent cardiovascular and ophthalmic diseases, and increase sap flow. Additionally, they serve as antipyretics. In traditional medicine, species like *D. huoshanense, D. nobile, D. chrysotoxum*, and *D. officinale* are known for their hypoglycemic and antioxidative properties, are considered commonly used for preparing tonics, regarded as superior to other *Dendrobium* species (Pan et al., 2014).

9.6 CONCLUSION

Orchids have been fascinating flowers for centuries to mankind for their varied forms, shades of color, longevity, and association with folklore in many cultures. The orchid flowers, besides being known for their elegance and beauty, also have use in the treatment of several ailments in traditional medicine systems across the globe, particularly in South Asian and South East Asian medicine systems of the Ayurveda, Chinese, Japanese, and in the Korean system, etc., for centuries. *Dendrobium* orchids have been widely used in Ayurveda, traditional Chinese medicine, and many other traditional medicine systems. Dendrobii Herba, Shi-Hu, Tie-Pi-Feng Dou that comprises many *Dendrobium* species such as *D. candidum/D. moniliforme, D. catenatum/D. officinale D. chrysanthum, D. chrysotoxum, D. nobile, D. fimbriatum var. oculatum, D. loddigessi*, etc. as have been used to treat bone fractures, burns, eye diseases, fever, kidney disorders, lung disorders, skin problem, gastritis, etc. The medicinal and therapeutic properties of any plant are due to the presence of bioactive compounds. Notably, the *Dendrobium* genus is known to contain numerous bioactive compounds, including flavonoids, phenylpropanoids, polysaccharides, bibenzyls, lignans, alkaloids, phenanthrenes, and others. These compounds are recognized for their pharmacological properties, such as antioxidant, anticancer, anti-inflammatory, immunomodulatory, antimicrobial, tyrosinase inhibitory activities, hypoglycemic, anti-aging, and anti-psoriasis. Thus, the *Dendrobium* species, alongside the traditional medicine system, can be explored for its application in the modern medicine system. Orchids have always been collected from the wild for the preparation of medicine, which is posing a threat to the species' population, rendering many extinct. Therefore, measures have to consider the commercialized cultivation of these species and prevent their collection from the wild as well as identify the species with high-quality bioactive compounds for isolation so their high pharmacological efficacy can be explored to treat dreaded diseases.

REFERENCES

Agoo, E. M. G. 2007. Status of orchid taxonomy research in the Philippines. *Journal of Systematic Biology* 1(1):1–25.

Akhter, M., Hoque, M. M., Rahman, M. & Huda, M. K. 2017. Ethnobotanical investigation of some orchids used by five communities of Cox's Bazar and Chittagong hill tracts districts of Bangladesh. *The Journal of Medicinal Plants* 5(3):265–268.

Arobaya, A. Y. S., Heipon, A., Sinery, A. S. & Wakum, A. 2020. Biodiversity mapping of *Dendrobium* Sw. section Spathulata Lindl. in the Ramsar Site of Wasur National Park of Indonesia New Guinea. *World Journal of Advanced Research and Reviews* 5(2):167–176.

Aung, Y. L., Mu, A. T., Aung, M. H., Liu, Q. & Jin, X. 2020. An annotated checklist of Myanmar orchid flora. In: Jin, X.-H., Xia, N.-H. & Tan, Y.-H. (Eds.), *Plant Diversity of Southeast Asia-II. Phyto Keys* 138:49–112.

Averyanov, L. V. & Averyanova, A. L. 2003. *Updated Checklist of the Orchids of Vietnam*. Hanoi: Vietnam National University Publishing House.

Awasthi, A. K. 1991. Ethnobotanical studies of the Negrito Islanders of Andaman Islands, India: The great Andamanese. *Economic Botany* 45(2):274–280.

Bhattacharya, M., Roy, A. K., Majumder, S., Ghosh, A., Chakraborty, S., Saha, S., Acharya, S. & Sarkar, S. 2020. Free radical scavenging activity versus flavonoid content in twelve *Dendrobium* orchids collected from Darjeeling Hills of Eastern Himalaya. *Species* 21(68):222–226.

Cakova, V., Bonte, F. & Lobstein, A. 2017. *Dendrobium*: Sources of active ingredients to treat age-related pathologies. *Aging and Disease* 8:827–849.

Chen, L., Wei, P. L., Zhou, H. B., Si, X. F., Mu, C. X. & Qiu, M. F. 2013. Studies and application of *Dendrobium* officinale. *Asia-Pacific Traditional Medicine* 9:51–53.

Chen, S. C. & Tsi, Z. H. 2000. *The Orchards of China*, 2nd ed. Beijing, China: The Chinese Forestry Press.

Chen, W., Lu, J., Zhang, J., Wu, J., Yu, L., Qin, L. & Zhu, B. 2021. Traditional uses, phytochemistry, pharmacology, and quality control of *Dendrobium* officinale Kimura et. Migo. *Frontiers in Pharmacology* 12:726528.

Chen, Y. G., Yu, H. & Lian, X. 2015. Isolation of stilbenoids and lignans from *Dendrobium* hongdie. *Tropical Journal of Pharmaceutical Research* 14(11):2055–2059.

Cheng, L. L., Yang, F. J., Wang, H. Y., Li, W. & Li, M. 2015. Latest research progress on *Dendrobium* devonianum. *Asia-Pacific Traditional Medicine* 11:31–33.

Chinese Pharmacopoeia. 2015. *Chinese Pharmacopoeia Commission*. Beijing, China: Chinese Medical Science and Technology Press, p. 1.

Choudhary, D., Mashkey, V. K., Goutam, E., Shrivastava, M., Rawat, M., Kumari, A. & Tripathi, V., 2023. Medicinal orchids: Traditional uses and recent advances. *Annals of Phytomedicine* 12(1):1–9.

Clements, M. A. 2003. Molecular phylogenetic systematics of the Dendrobiinae (Orchidaceae), with emphasis on *Dendrobium* section *Pedilonum*. *Telopea* 10(1):247–298.

De, L. C., Rao, A. N., Rajeevan, P. K., Pathak, P. & Singh, D. R. 2015a. Medicinal and aromatic orchids: An overview. *International Journal of Current Research* 7(9):19931–19935.

De, L. C., Rao, A. N., Rajeevan, P. K., Srivastava, M. & Chhetri, G. 2015b. Morphological characterization in *Dendrobium* species. *Journal of Global Biosciences* 4(1):1198–1215.

Deori, C., Sarma, S. K. & Hynniewta, T. M. 2019. *Dendrobium Orchids of Northeast India*. Guwahati, Assam: Purbayon Publication, p. 407.

Devaraj, R. D., Reddy, C. K. & Xu, B. 2019. Health-promoting effects of konjac glucomannan and its practical applications: A critical review. *International Journal of Biological Macromolecules* 129:273–281.

Dizon, S. A., Ocenar, A. P. & Naïve, M. A. K. 2018. Inventory of orchids in the Mount Hamiguitan Range Wildlife Sanctuary, Davao Oriental, Philippines. *Bio Bulletin* 4(1):37–42.

Duong, T. D., Trung, K. H., Nghia, L. T., Thuy, N. T., Hien, P. B., Khoa, N. T., Dũng, T. H., Trung, D. M. & Khanh, T. D. 2018. Identification of Vietnamese native *Dendrobium* species based on ribosomal DNA internal transcribed spacer sequence. *Advanced Studies in Biology* 10:1–12.

Duraipandiyan, V., Ayyanar, M. & Ignacimuthu, S. 2006. Antimicrobial activity of some ethnomedicinal plants used by Paliyar tribe from Tamil Nadu, India. *BMC Complementary Alternative Medicine* 17(6):35.

Fan, C., Wang, W., Wang, Y., Qin, G. & Zhao, W. 2001. Chemical constituents from *Dendrobium densiflorum. Phytochemistry* 57(8):1255–1258.

Fernando, S. S. & Ormerod, P. 2008. An annotated checklist of the orchids of Sri Lanka. *Rheedea* 18(1):1–28.

Fischer, G. A., Gravendeel, B., Sieder, A., Andriantiana, J., Heiselmayer, P., Cribb, P. J., de Camargo Smidt, E., Samuel, R. & Kiehn, M. 2007. Evolution of resupination in Malagasy species of Bulbophyllum (Orchidaceae). *Molecular Phylogenetics and Evolution* 45:358–376.

Ganapaty, S., Ramaiah, M., Yasaswini, K., Nuthakki, V. K. & Harikrishnareddy, D. 2013. Quantitative phytochemical estimation and evaluation of hepatoprotective activity of methanolic extract of *Dendrobium ovatum* (L.) Kraenzl. whole plant against CCl4 induced hepatotoxicity. *Journal of Pharmacognosy and Phytochemistry* 2(3):113–118.

Gao, L., Wang, F., Hou, T., Geng, C., Xu, T., Han, B. & Liu, D. 2022. *Dendrobium huoshanense* CZ Tang et SJ Cheng: A review of its traditional uses, phytochemistry, and pharmacology. *Frontiers in Pharmacology* 13:920823.

Go, R., Eng, K. H., Mustafa, M., Abdullah, J. O., Naruddin, A. A., Lee, N. S., Lee, C. S., Eum, S. M., Park, K. W. & Choi, K. 2011. An assessment of orchids' diversity in Penang Hill, Penang, Malaysia after 115 years. *Biodiversity and Conservation* 20:2263–2272.

Gong, C. Y., Yu, Z. Y., Lu, B., Yang, L., Sheng, Y. C., Fan, Y. M. & Wang, Z. T. 2014. Ethanol extract of *Dendrobium* chrysotoxum Lindl ameliorates diabetic retinopathy and its mechanism. *Vascular Pharmacology* 62(3):134–142.

Gu, C., Zhang, X., Wu, L., Jiang, X. & Huang, L. 2017. Quality evaluation of *Dendrobium* based on ultra-performance liquid chromatography (UPLC) and chemometrics. *Journal of Applied Pharmaceutical Science* 7(1):17–23.

Haizhen, B., Shengrong, S., Yimin, Z., Sun, S. & Zhongwei, C. 2017. Inhibitory effect of *Dendrobium officinale* polysaccharide on growth of human breast cancer MCF-7 cells and the related mechanism. *Food Science Technology* 28:1922–1926.

Hartati, S., Samanhudi, S. & Cahyono, O. 2022. Morphological characterization of five species of *Dendrobium* native to Indonesia for parent selection. *Biodiversitas Journal of Biological Diversity* 23(5):2648–2654.

Hegde, S. N., Tam, N., Tsering, J. & Apang, O. 2017. *Orchids of Aruanchal Pradesh.* Department of Environment and Forests, Government of Arunachal Pradesh, p. 168.

Hossain, M. M. 2009. Traditional therapeutic uses of some indigenous orchids of Bangladesh. *Medicinal and Aromatic Plant Science and Biotechnology* 42(1):101–106.

Hou, S. Z., Liang, C. Y., Liu, H. Z., Zhu, D. M., Wu, Y. Y., Liang, J. & Lai, X. P. 2016. *Dendrobium officinale* prevents early complications in streptozotocin-induced diabetic rats. *Evidence-Based Complementary and Alternative Medicine* 2016:6385850.

Htwe, A. N. 2019. Taxonomic study on five *Dendrobium* species from Mt. Poppa area. *Yadanabon University Research Journal* 10(1):1–8.

Hu, Y., Ren, J., Wang, L., Zhao, X., Zhang, M., Shimizu, K. & Zhang, C. 2018. Protective effects of total alkaloids from *Dendrobium* crepidatum against LPS-induced acute lung injury in mice and its chemical components. *Phytochemistry* 149:12–23.

Hwang, J. S., Lee, S. A., Hong, S. S., Han, X. H., Lee, C., Kang, S. J., Lee, D., Kim, Y., Hong, J. T., Lee, M. K. & Hwang, B. Y. 2010. Phenanthrenes from *Dendrobium nobile* and their inhibition of the LPS-induced production of nitric oxide in macrophage RAW 264.7 cells. *Bioorganic & Medicinal Chemistry Letters* 20(12):3785–3787.

Isa, S. F. M., Seok Yien Yong, C., Abdullah, J. O. & Go, R. 2018. Annotated checklist of orchids found in Merapoh Trail (Gunung Tahan, Malaysia). *Pertanika Journal of Tropical Agricultural Science* 41(3):1245–1259.

Jalal, J. S., Kumar, P. & Pangtey, Y. P. S. 2008. Ethnomedicinal orchids of Uttarakhand, Western Himalaya. *Ethnobotanical Leaflets* 1:164.

Jin, X. H. & Huang, L. Q. 2015. Investigation of original materials of Chinese medicine "Shihu" and "Tiepishihu". *Zhongguo Zhong yao za zhi = Zhongguo Zhongyao Zazhi = China Journal of Chinese Materia Medica* 40(13):2475–2479.

Jones, K. 2018. *The Genus Dendrobium.* Western Australia: The Species Orchid Society of WA, p. 63.

Joshi, P. R., Paudel, M. R., Chand, M. B., Pradhan, S., Pant, K. K., Joshi, G. P., Bohara, M., Wagner, S. H., Pant, B. & Pant, B. 2020. Cytotoxic effect of selected wild orchids on two different human cancer cell lines. *Heliyon* 6(5):e03991.

Kim, J. H., Oh, S. Y., Han, S. B., Uddin, G. M., Kim, C. Y., & Lee, J. K. 2015. Anti-inflammatory effects of *Dendrobium nobile* derived phenanthrenes in LPS-stimulated murine macrophages. *Archives of Pharmacal Research* 38:1117–1126.

Klongkumnuankarn, P., Busaranon, K., Chanvorachote, P., Sritularak, B., Jongbunprasert, V. & Likhitwitayawuid, K. 2015. Cytotoxic and antimigratory activities of phenolic compounds from *Dendrobium brymerianum*. *Evidence-Based Complementary and Alternative Medicine* 2015(1):350410.

Kovács, A., Vasas, A. & Hohmann, J. 2008. Natural phenanthrenes and their biological activity. *Phytochemistry* 69(5):1084–1110.

Lam, Y., Ng, T. B., Yao, R. M., Shi, J., Xu, K., Sze, S. C. W. & Zhang, K. Y. 2015. Evaluation of chemical constituents and important mechanism of pharmacological biology in *Dendrobium* plants. *Evidence-Based Complementary and Alternative Medicine* 2015:841752.

Li, C. H. & Li, K. C. 2017. Propagation and greenhouse production of *Dendrobium*. *China Flowers Horticulture* 16:27–31.

Li, J., Zhang, X., Li, Y., Liu, B. & Zhan, R. 2023. Bibenzyls and lignans from *Dendrobium devonianum* and their immunosuppressive activities. *Phytochemistry Letters* 54:137–140.

Li, P. Y., Li, L. & Wang, Y. Z. 2023. Traditional uses, chemical compositions and pharmacological activities of *Dendrobium*: A review. *Journal of Ethnopharmacology* 310:116382.

Li, Q., Chen, J., Li, Y., Chen, T., Zou, J. & Wang, H. 2017. Effect of polysaccharide of *Dendrobium candidum* on proliferation and apoptosis of human corneal epithelial cells in high glucose. *Medicine* 96(32):e7773.

Li, W. H. 1995. *Medical Records, in Compilation Committee of Local Records, in Records of Yunnan Province*. Kunming: Yunnan People's Publishing House, pp. 960–962.

Li, X. W., Huang, M., Lo, K., Chen, W. L., He, Y. Y., Xu, Y., Zheng, H., Hu, H. & Wang, J. 2019. Anti-diabetic effect of a Shihunine-rich extract of *Dendrobium loddigesii* on 3T3-L1 cells and db/db mice by up-regulating AMPK–GLUT4–PPARα. *Molecules* 24:2673.

Li, Y. M., Wang, H. Y. & Liu, G. Q. 2001. Erianin induces apoptosis in human leukemia HL-60 cells. *Acta Pharmacologica Sinica* 22(11):1018–1022.

Limpanit, R., Chuanasa, T., Likhitwitayawuid, K., Jongbunprasert, V. & Sritularak, B. 2016. a-Glucosidase inhibitors from *Dendrobium* tortile. *Records of Natural Products* 10(5):609.

Lin, C. F., Lay, H. L., Ni, C. L. & Chen, C. C. 2013. Phenolic components of *Dendrobium antennatum*. *Chemistry of Natural Compounds* 49:520–522.

Lin, Y., Wang, F., Yang, L. J., Chun, Z., Bao, J. K. & Zhang, G. L. 2013. Anti-inflammatory phenanthrene derivatives from stems of *Dendrobium denneanum*. *Phytochemistry* 95:242–251.

Liu, G. Y., Tan, L., Cheng, L., Ding, L. S., Zhou, Y., Deng, Y., He, Y. Q., Guo, D. L. & Xiao, S. J. 2020. Dendrobine-type alkaloids and bibenzyl derivatives from *Dendrobium findlayanum*. *Fitoterapia* 142:104497.

Liu, Q., Zhou, S. S., Jin, X. H., Pan, B., Maung, K. W., Zyaw, M., Li, R., Quan, R.-C. & Tan, Y.-H. 2018. *Dendrobium naungmungense* (Orchidaceae, Dendrobieae), a new species from Kachin State, Myanmar. *Phyto Keys* 94:31–38.

Lokho, A. 2013. Diversity of *Dendrobium* Sw. its distributional patterns and present status in the Northeast India. *International Journal of Scientific and Research Publications* 3(5):1–9.

Longchar, T. B. & Deb, C. R. 2021. Comparative analysis of nutraceutical potential phytochemicals and antioxidant activities in different parts of wild and in vitro regenerated plantlets of *Dendrobium heterocarpum* Wall. ex Lindl.: A medicinal orchid. *Journal of Pharmacognosy and Phytochemistry* 10(4):331–336.

Meng, Q., Chen, F., Xiao, T. & Zhang, L. 2019. Superfine grinding of *Dendrobium officinale*: The finer the better? *International Journal of Food Science and Technology* 54:2199–2208.

Millar, A. 1978. *Orchids of Papua New Guinea an Introduction*. Canberra: Australian National University Press.

Moudi, M. & Go, R. 2015. Monophyly of four sections of genus *Dendrobium* (Orchidaceae): Evidence from nuclear ribosomal DNA Intrenal Transcribed Spacer (ITS) sequences. *International Journal of Bioassays* 4(1):3622–3626.

Ninawe, A. S. & Swapna, T. S. 2017. Orchid diversity of Northeast India–traditional knowledge and strategic plan for conservation. *Journal of Orchid Society of India* 31:41–56.

Pan, C., Chen, S., Chen, Z., Li, Y., Liu, Y., Zhang, Z., Xu, Y., Liu, G., Yang, K., Liu, G. & Du, Z. 2022. Assessing the geographical distribution of 76 *Dendrobium* species and impacts of climate change on their potential suitable distribution area in China. *Environmental Science and Pollution Research* 29(14):20571–20592.

Pan, L. H., Feng, B. J., Wang, J. H., Zha, X. Q. & Luo, J. P. 2013. Structural characterization and anti-glycation activity in vitro of a water-soluble polysaccharide from *Dendrobium huoshanense*. *Journal of Food Biochemistry* 37(3):313–321.

Pan, L. H., Li, X. F., Wang, M. N., Zha, X. Q., Yang, X. F., Liu, Z. J., Luo, Y. B. & Luo, J. P. 2014. Comparison of hypoglycemic and antioxidative effects of polysaccharides from four different *Dendrobium* species. *International Journal of Biological Macromolecules* 64:420–427.

Pant, B. 2013. Medicinal orchids and their uses: Tissue culture a potential alternative for conservation. *African Journal of Plant Science* 7(10):448–467.

Pant, B., Chand, K., Paudel, M. R., Joshi, P. R., Thapa, B. B., Park, S. Y., Shakya, S., Thakuri, L. S., Rajbahak, S., Sah, A. K. & Baniya, M. K. 2022. Micropropagation, antioxidant and anticancer activity of pineapple orchid: *Dendrobium densiflorum* Lindl. *Journal of Plant Biochemistry and Biotechnology* 31:399–409.

Paudel, M. R., Chand, M. B., Pant, B. & Pant, B. 2019. Assessment of antioxidant and cytotoxic activities of extracts of *Dendrobium* crepidatum. *Biomolecules* 9(9):478.

Paudel, M. R., Joshib, P. R., Chandb, K., Sahb, A. K., Acharyab, S., Pant, B. & Pant, B. 2020. Antioxidant, anticancer and antimicrobial effects of In vitro developed protocorms of *Dendrobium longicornu*. *Biotechnology Reports* 28:e00527.

Peyachoknagul, S., Mongkolsiriwatanab, C., Wannapinpong, S., Huehnea, P. S. & Srikulnath, K. 2014. Identification of native *Dendrobium* species in Thailand by PCR-RFLP of rDNA-ITS and chloroplast DNA. *Science Asia* 40:113–120.

Qi, L., Shi, Y., Li, C., Liu, J., Chong, S. L., Lim, K. J., Si, J., Han, Z. & Chen, D. 2022. Glucomannan in *Dendrobium catenatum*: Bioactivities, biosynthesis and perspective. *Genes* 13(11):1957.

Rahamtulla, M., Mallikarjuna, K. & Khasim, S. M. 2023. GC-MS analysis and therapeutic importance of leaf extracts of *Dendrobium aphyllum* (Roxb.) CEC Fischer: An *In vitro* study. *South African Journal of Botany* 153:62–76.

Rahamtulla, M., Pradhan, U. C., Roy, A. K., Rampilla, V. & Khasim, S. M. 2020. Ethnomedicinal aspects of some orchids from Darjeeling Himalaya, India. In: *Orchid Biology: Recent Trends & Challenges*. Berlin, Germany: Springer, pp. 451–472.

Rahman, M. & Huda, M. K. 2021. Exploration of phytochemical, antioxidant and anti-inflammatory efficacy of the ethnomedicinal uses of ten orchids of Bangladesh. *Advancement in Medicinal Plant Research* 9(2):30–39.

Ramesh, T., Koperuncholan, M., Praveena, R., Ganeshkumari, K., Vanithamani, J., Muruganantham, P. & Renganathan, P. 2019. Medicinal properties of some *Dendrobium* orchids: A review. *Journal of Applied and Advanced Research* 4(4):119–128.

Rao, N. Y. & Babu, N. K. 2017. Phytochemical analysis and acute toxicity studies of methanolic extract of stem of *Dalbergia lanceolaria*, flowers of *Dendrobium normale* and bark of *Measa indica*. *Research Journal of Pharmaceutical Biological and Chemical Sciences* 8(2):97–100.

Rao, R. R. 1981. Medicinal plants used by Khasi and Garo tribes. *Economic Botany* 35(1):4–9.

Rokaya, M. B., Raskoti, B. B., Timsina, B. & Münzbergová, Z. 2013. An annotated checklist of the orchids of Nepal. *Nordic Journal of Botany* 31:511–550.

Saavedra, A. J. L. & Pitogo, K. M. E. 2020. Richness and distribution of orchids (Orchidaceae) in the forests of Mount Busa, Sarangani, Southern Mindanao, Philippines. *Philippine Journal of Science* 150(S1):151–163.

Sandamali, P. M. H., Senanayake, S. P. & Rajapakse, S. 2020. Phylogenetic relationships of selected Sri Lankan Orchids based on Internal Transcribed Spacer (ITS) sequence analysis. *Tropical Plant Research* 7(1):76–85.

Schuiteman, A., Bonnet, P., Svengsuksa, B. & Barthélémy, D. 2008. An annotated checklist of the Orchidaceae of Laos. *Nordic Journal of Botany* 26:257–316.

Shang, Z. M. 2019. Chemical constituents of *Dendrobium bellatulum*. *Chinese Traditional and Herbal Drugs* 2019:2036–2040.

Shankar, R., Lavekar, G. S., Deb, S. & Sharma, B. K. 2012. Traditional healing practice and folk medicines used by Mishing community of North East India. *Journal of Ayurveda and Integrative Medicine* 3(3):124.

Sharma, U., Rao, R. V., Mohan, J. S. S. & Reddy, A. S. 2007. In vitro propagation of *Dendrobium microbulbon* A. Rich – A rare ethnomedicinal herb. *Indian Journal of Biotechnology* 6:381–384.

Singh, A. & Duggal, S. 2009. Medicinal orchids: An overview. *Ethnobotanical Leaflets* 3:3.

Singh, B. 2022. Therapeutic Himalayan herbs: Folklore uses, bioactive phytochemicals, and biological activities of medicinal orchids used by Nomads. *Indian Journal of Natural Products and Resources (IJNPR) [Formerly Natural Product Radiance: (NPR)]* 13(1):94–104.

Singh, C. B., Devi, M. C., Thokchom, D. S., Sengupta, M. & Singh, A. K. 2015. Phytochemical screening, estimation of total phenols, total flavonoids and determination of antioxidant activity in the methanol extract of *Dendrobium denudans* D. Don stems. *Journal of Pharmacognosy and Phytochemistry* 4(4):6–11.

Singh, K. P., Phukan, S. & Bujarbarua, P. 2001. *Floristic Diversity and Conservation Strategies in India.* Botanical Survey of India, IV:1736–1764.

Singh, S., Singh, A. K., Kumar, S., Kumar, M., Pandey, P. K. & Singh, M. C. K. 2012. Medicinal properties and uses of orchids: A concise review. *Elixir Applied Botany* 52:11627–11634.

Sliwinski, T., Kowalczyk, T., Sitarek, P. & Kolanowska, M. 2022. Orchidaceae-derived anticancer agents: A review. *Cancers* 14:754.

Smith, N. M. 1991. Ethnobotanical field notes from the Northern Territory, Australia. *Journal of the Adelaide Botanic Garden* 14(1):1–65.

Sritularak, B. & Likhitwitayawuid, K. 2009. New bisbibenzyls from *Dendrobium* falconeri. *Helvetica Chimica Acta* 92(4):740–744.

Sukphan, P., Sritularak, B., Mekboonsonglarp, W., Lipipun, V. & Likhitwitayawuid, K. 2014. Chemical constituents of *Dendrobium venustum* and their antimalarial and anti-herpetic properties. *Natural Product Communications* 9(6):1934578X1400900625.

Takamiya, T., Wongsawad, P., Tajima, N., Shioda, N., Lu, J. F., Wen, C. L. & Yukawa, T. 2011. Identification of *Dendrobium* species used for herbal medicines based on ribosomal DNA internal transcribed spacer sequence. *Biological and Pharmaceutical Bulletin* 34(5):779–782.

Tanagornmeatar, K., Chaotham, C., Sritularak, B., Likhitwitayawuid, K. & Chanvorachote, P. 2014. Cytotoxic and anti-metastatic activities of phenolic compounds from *Dendrobium* ellipsophyllum. *Anticancer Research* 34(11):6573–6579.

Tsering, J., Tam, N., Tag, H., Gogoi, B. J. & Apang, O. 2017. Medicinal orchids of Arunachal Pradesh: A review. *Bulletin of Arunachal Forest Research* 32(1&2):1–16.

Tucci, J. & Wilkens, S. 2016. A brief review of the application and pharmacology of ethnomedicines of Indigenous Australians. *Australian Journal of Rural Health* 24(3):156–169.

Vinayagam, R., Jayachandran, M. & Xu, B. 2016. Antidiabetic effects of simple phenolic acids: A comprehensive review. *Phytotherapy Research* 30(2):184–199.

Wagh, V. V. & Jain, A. K. 2020. Ethnopharmacological survey of plants used by the Bhil and Bhilala ethnic community in dermatological disorders in Western Madhya Pradesh, India. *Journal of Herbal Medicine* 19:100234.

Wang, Y. H. 2021. Traditional uses and pharmacologically active constituents of *Dendrobium* plants for dermatological disorders: A review. *Natural Products and Bioprospecting* 11:465–487.

Warinhomhoun, S., Khine, H. E. E., Sritularak, B., Likhitwitayawuid, K., Miyamoto, T., Tanaka, C., Punsawad, C., Punpreuk, Y., Sungthong, R. & Chaotham, C. 2022. Secondary metabolites in the *Dendrobium heterocarpum* methanolic extract and their impacts on viability and lipid storage of 3T3-L1 pre-adipocytes. *Nutrients* 14(14):2886.

Wei, X., Chen, D., Yi, Y., Qi, H., Gao, X., Fang, H. & Gu, L. 2012. Syringic acid extracted from *Herba dendrobii* prevents diabetic cataract pathogenesis by inhibiting aldose reductase activity. *Evidence-Based Complementary and Alternative Medicine* 2012(1):426537.

Wongjuk, J. & Thirabunyanon, M. 2020. Bioactivity of crude extract from *Dendrobium* spp. on antioxidant and colon cancer cell antiproliferation (Doctoral dissertation, Maejo University), 114 p.

Wood, H. P. 2006. *The Dendrobiums.* Germany: A. R. G. Gantner Verlag K. G.

Xing, S., Yu, W., Zhang, X., Luo, Y., Lei, Z., Huang, D., Lin, J., Huang, Y., Huang, S., Nong, F. & Zhou, C. 2018. Isoviolanthin extracted from *Dendrobium officinale* reverses TGF-β1-mediated epithelial-mesenchymal transition in hepatocellular carcinoma cells via deactivating the TGF-β/Smad and PI3K/Akt/mTOR signaling pathways. *International Journal of Molecular Sciences* 19:1556.

Xiong, L., Cao, Z. X., Peng, C., Li, X. H., Xie, X. F., Zhang, T. M. & Guo, L. 2013. Phenolic glucosides from *Dendrobium aurantiacum* var. denneanum and their bioactivities. *Molecules* 18(6):6153–6160.

Xu, J., Han, Q. B., Li, S. L., Chen, X. J., Wang, X. N., Zhao, Z. Z. & Chen, H. B. 2013a. Chemistry, bioactivity and quality control of *Dendrobium*, a commonly used tonic herb in traditional Chinese medicine. *Phytochemistry Reviews* 12:341–367.

Xu, J., Li, S. L., Yue, R. Q., Ko, C. H., Hu, J. M., Liu, J., Ho, H. M., Yi, T., Zhao, Z. Z., Zhou, J., Leung, P. C., Chen, H. B. & Han, Q. B. 2014. A novel and rapid HPGPC-based strategy for quality control of saccharide-dominant herbal materials: *Dendrobium officinale*, a case study. *Analytical and Bioanalytical Chemistry* 406(25):6409–6417.

Xu, Q., Liu, Y., Cong, Y. B., Zheng, Y. Y., Zhang, J. P., Cui, Y. & Shi, H. 2013b. Gene expression and microarray investigation of *dendrobium* mixture as progressive therapy for the treatment of type 2 diabetes mellitus. *Tropical Journal of Pharmaceutical Research* 12(2):195–201.

Yang, D., Liu, L. Y., Cheng, Z. Q., Xu, F. Q., Fan, W. W., Zi, C. T. & Hu, J. M. 2015. Five new phenolic compounds from *Dendrobium aphyllum. Fitoterapia* 100:11–18.

Yang, H., Sung, S. H. & Kim, Y. C. 2007. Antifibrotic phenanthrenes of *Dendrobium nobile* stems. *Journal of Natural Products* 70(12):1925–1929.

Yang, L., Qin, L. H., Bligh, S. A., Bashall, A., Zhang, C. F., Zhang, M. & Xu, L. S. 2006. A new phenanthrene with a spirolactone from *Dendrobium chrysanthum* and its anti-inflammatory activities. *Bioorganic & Medicinal Chemistry* 14(10):3496–3501.

Yang, L., Wang, Z. & Xu, L. 2006. Phenols and a triterpene from *Dendrobium aurantiacum* var. denneanum (Orchidaceae). *Biochemical Systematics and Ecology* 8(34):658–660.

Yang, Y.-J., Tan, S.-L., Huang, J.-L., Zhang, S.-B. & Huang, W. 2020. The water-water cycle facilitates photosynthetic regulation under fluctuating light in the epiphytic orchid *Dendrobium officinale. Environmental and Experimental Botany* 180:104238.

Ye, Z., Dai, J. R., Zhang, C. G., Lu, Y., Wu, L. L., Gong, A. G. & Wang, Z. T. 2017. Chemical differentiation of *Dendrobium officinale* and *Dendrobium devonianum* by using HPLC fingerprints, HPLC-ESI-MS, and HPTLC analyses. *Evidence-Based Complementary and Alternative Medicine* 2017(1):86.

Yoo, S. R., Jeong, S. J., Lee, N. R., Shin, H. K. & Seo, C. S. 2017. Simultaneous determination and anti-inflammatory effects of four phenolic compounds in Dendrobii Herba. *Natural Product Research* 31(24):2923–2926.

Yu, Z., Gong, C., Lu, B., Yang, L., Sheng, Y., Ji, L. & Wang, Z. 2015. *Dendrobium chrysotoxum* Lindl. alleviates diabetic retinopathy by preventing retinal inflammation and tight junction protein decrease. *Journal of Diabetes Research* 2015(1):518317.

Zhang, C. F., Wang, M., Wang, L., Iinuma, M., Zhang, M., Xu, L. S. & Wang, Z. T. 2008. Chemical constituents of *Dendrobium gratiosissimum* and their cytotoxic activities. *Indian Journal of Chemistry* 47B:952–956.

Zhang, H. Y. & Zhang, Z. Y. 1994. *Handbook of Traditional Chinese Medicine Resources in China.* Beijing: Science Press, pp. 1531–1536.

Zhang, X., Xu, J. K., Wang, N. L., Hiroshi, K., Yao, X. S. & Wang, Z. 2008. Studies on antioxidant activity of bibenzyls and phenolic components from *Dendrobium nobile. Chinese Pharmaceutical Journal* 43(11):829–832.

Zhang, Z., Zhang, D., Dou, M., Li, Z., Zhang, J. & Zhao, X. 2016. *Dendrobium officinale* Kimura et Migo attenuates diabetic cardiomyopathy through inhibiting oxidative stress, inflammation and fibrosis in streptozotocin-induced mice. *Biomedicine & Pharmacotherapy* 84:1350–1358.

Zhang, Z. M., Liu, S., Yang, H., Wang, N., Zou, Y. H., Zhuang, P. Y. & Liu, H. 2022. Chemical constituents from *Dendrobium chrysanthum* and their chemotaxonomic significance. *Biochemical Systematics and Ecology* 105:104522.

Zhong, K. & Yang, T. 2021. Study on HPLC fingerprint of *Dendrobium nobile* Lindl. Flower based on chemometrics. *IOP Conference Series: Earth and Environmental Science* 651:032027.

Zhou, X. M., Zheng, C. J., Gan, L. S., Chen, G. Y., Zhang, X. P., Song, X. P. & Sun, C. G. 2016. Bioactive phenanthrene and bibenzyl derivatives from the stems of *Dendrobium nobile. Journal of Natural Products* 79(7):1791–1797.

10 Medicinal Orchids of Genus *Habenaria*

Noyaljohnson Tinoammini, Gopalakrishnan Aazhivaendhan, Chinnaiyan Rajasekar, Govindarajan Siva, Thiruppathi Senthil Kumar, and Mariappan Muthukumar

10.1 INTRODUCTION

Phytotherapeutic medications have gained popularity as an effective and safer holistic approach to disease management (Girija and Sivan, 2022). The demand for herbal medicine began to increase gradually after the pandemic due to its effectiveness against antibiotic-resistant microbes. Numerous bioactive compounds have been isolated from plants and studied for their biomedical application every year. Orchids are an exceptional group of flowering plants renowned for their aesthetic cut-flowers in floriculture and their effective use in folk medicine. The advances in orchid research paved the way for the development of brand-new products in cosmetic industries. Exploring medicinal orchids and learning about their traditional remedies helps the pharmaceutical sector to produce more plant-derived natural products (Eldin et al., 2023). Orchids are rich in phytochemical constituents such as alkaloids, flavonoids, glycosides, carotenoids, saponins, xanthones, and carbohydrates, which were employed in traditional systems of medicine (Jalal et al., 2008; Hossain, 2011). Various indigenous practices such as Ayurveda, Homeopathy, Folk, Complementary, and Chinese traditional medicine have documented the use of orchids in their formulations (Liu et al., 2014; Kanlayavattanakul and Lourith, 2022). Orchids exhibit two major forms: terrestrial orchids and epiphytic orchids. Both forms have been reported for their therapeutic value since the time immemorial. The Epiphytic genera having medicinal properties are *Dendrobium, Eria, Luisia*, and *Vanda*, followed by terrestrial orchids such as *Anoectochilus, Calanthe, Coelogyne, Cymbidium, Cypripedium, Ephemerantha, Gastrodia, Gymnadenia, Habenaria, Ludisia, Malaxis, Orchis*, and *Thunia* (Bungtongdee et al., 2019; Li et al., 2023). Terrestrial orchids possess a more prominent curative ability compared to their epiphytic counterparts due to ecological diversity, evolutionary changes, and mycorrhizal fungi, which influence their synthesis of various secondary metabolites.

Habenaria is a peculiar group of orchids distributed across the globe, and it has more than 844 distinct species. These orchids are widely recognized as "rein orchids" or "bog orchids" and are regarded as the largest terrestrial genus in Orchidaceae. The characteristic features to identify this genus are the presence of viscidium at the spur opening, unfused anther canals, bifid petals, and intensely divided labellum (Pridgeon et al., 2001). *Habenaria* flowers are especially known for their mimicry mechanism aimed at attracting pollinators (Chen et al., 2021). It also captures the attention of horticulturists and researchers for their ornamental value. The primary objective is to provide a comprehensive summary of the available information on the genus *Habenaria,* including food practices, therapeutic values, ethnopharmacology, and their bioactive compounds.

10.2 HABENARIA: CONCEALED TREASURE IN TRADITIONAL HEALING

Habenaria holds a significant place in the conventional system of medicine because of its ability to balance Vata and Pitta doshas. The native inhabitants of Ethiopia consume *Habenaria* species to

 DOI: 10.1201/b23375-10

overcome impotence. India and China are among the top countries famous for herbal medicine. As of now, India has documented 61 *Habenaria* species, among which 17 species are acknowledged for their medicinal properties. China reports 54 species, and 13 species have been recognized for traditional uses (Hu et al., 2000; Chen and Cribb, 2009; Singh et al., 2019).

Kasthausadhi is an Ayurvedic concoction purely made up of medicinal plants. Similarly, the plants are collected, processed, and supplied in the forms of Aristha (decoction), Avaleha (semi-solid decoction), Churana (powder), Ghrita (medicated ghee), and Taila (oil). Various parts of the *Habenaria* orchid, such as a whole plant or particular parts like the leaf, stem, flower, and tuber, are derived for several medications (Figure 10.1). Astavarga is a Kasthausadhi that is reported for immune boosters, cell restoration, unusual thirst, and diabetic mellitus. This special rejuvenating formulation is composed of eight herbs, of which two *Habenaria* species, *H. edgeworthii,* and *H. intermedia* are utilized. The Ayurvedic text mentions them in the Sanskrit names Riddhi and Vriddhi correspond to *H. edgeworthii* and *H. intermedia*, respectively (Gupta et al., 2023; Qiu et al., 2023). Tribal usage of *Habenaria* was documented in Siddha to reduce scrotum swellings by the pulse diagnosticians of Kolli Hills from Eastern Ghats (Subramani and Goraya, 2003) and for the treatment of syncope and hernia in Unani medicine. Chinese traditional partitioners also make use of *Habenaria* in preparation for anti-tuberculin medicine and for a broad range of ailments such as kidney-related disorders, stomach aches, sexually transmitted diseases, and so on (Chen and Tang, 1982; Lihua, 2015).

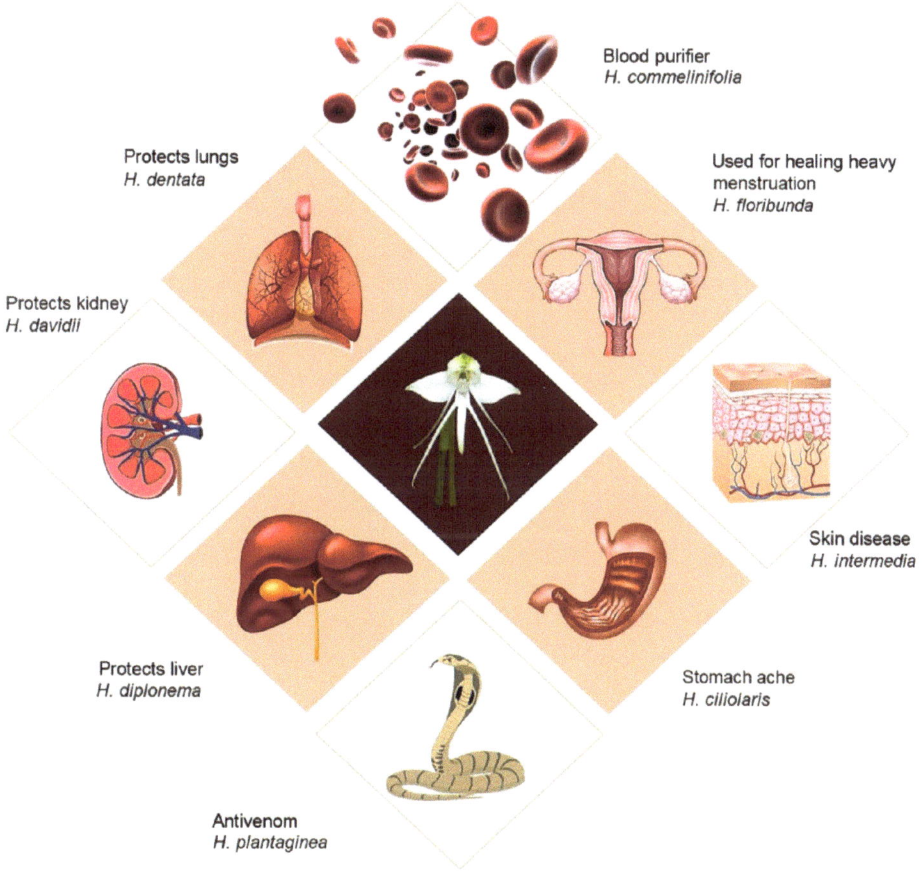

FIGURE 10.1 Therapeutic profile of *Habenaria* for healing various ailments.

10.3 CULTURAL FOOD PRACTICES

Plants exhibit a wide range of bioactive phytochemicals that have potential health benefits when included in our diet. Chikanda, a protein-rich food made with the tubers of two *Habenaria* species, is commonly used by Zambians and Bayam people in Cameroon (Menzepoh, 2011). *H. rumphii* tubers are utilized in preparing candies and jams and offer a promising avenue for culinary applications across various cuisines (Nurfadilah, 2020). Likewise, the edible tubers of *H. multipartita, H. malintana, H. commelinifolia*, and *H. marginata* are ingested as food items (Davis and Steiner, 1982; Kumar, 2003). The indigenous communities in Northeastern India rely on *H. susannae* as a traditional beverage for promoting health and well-being (Medhi and Chakrabarti, 2009).

10.4 PHYTOCHEMICALS

Plants produce bioactive chemical compounds from both primary and secondary metabolism, offering valuable health benefits. Identifying and understanding the phytochemicals in orchids that tribal communities use for treating various diseases is of utmost importance. In this context, we would like to highlight the phytochemicals that are found in the *Habenaria* genus (Figure 10.2). *H. intermedia* used in ayurveda preparation is composed of thiamine, tannins, phenol, and a higher amount of calcium. Similarly, *H. egdeworthii* is abundant in sodium and contains hydroxybenzoic acid and gallic acid (Rawat et al., 2014). The phenolic compounds, including habenarioside, loroglossin, militarin, and flavonoids like isoquercitrin and isorhamnetin 3,7-di-O-β-D-glucopyranoside, were also extracted from *H. petalodes* (Cota et al., 2008). Kumar and his associates (2023) isolated six compounds, namely, batatasin III, linoleic acid, loroglossol, β-sitosterol, gymconopin A, HBR-6 (3,5-Dimethoxy-3'-hydroxy Bibenzyl), and HBR-7 (3,5'-Dihydroxy-2(p-hydroxybenzyl)-3-methoxy Bibenzyl) from *H. intermedia*.

10.5 ETHNOPHARMACOLOGY

Ethnopharmacology is an interdisciplinary study of economic botany, anthropology, and pharmacology, as well as the traditional use of medicinal plants for therapeutic purposes. The indigenous communities possess extensive knowledge regarding the healing qualities of each species. Herbal

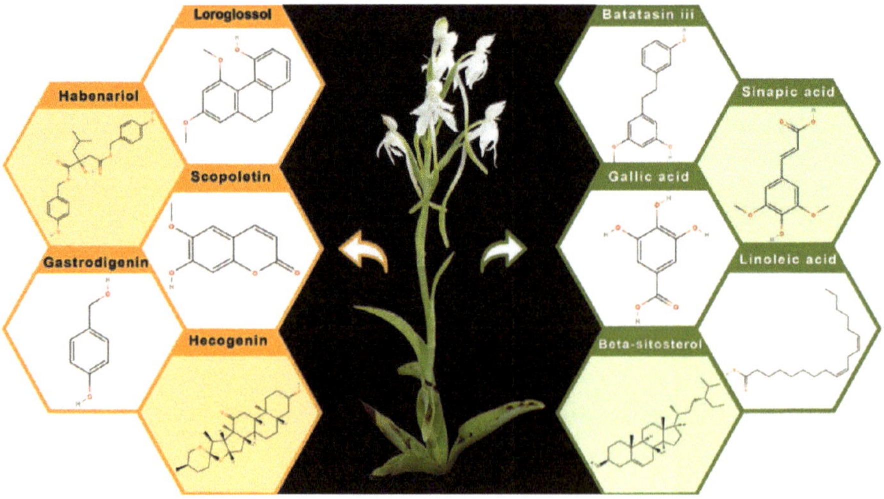

FIGURE 10.2 Bioactive phytochemicals in the *Habenaria* genus.

therapies employed by indigenous communities have proven to be invaluable reservoirs of inspiration for developing pharmaceutical drugs. Likewise, investigating the traditional applications of *Habenaria* orchids in addressing various health issues offers a promising avenue for uncovering novel chemical compounds in the realm of drug discovery with potential therapeutic benefits.

10.5.1 Hernia

Protruding of an organ or tissue through an abnormal opening in the fascia is known as hernia. *H. ciliolaris* is used to treat hernia, followed by *H. davidii* and *H. petelotii* in decoction form for inflammations in lymph nodes. Chinese herbal medicine reports *H. petelotii* and *H. stenopetala* are used to treat external bleeding injuries, sexual functions, and nocturnal bed-wetting in children. *H. delavayi,* believed to strengthen the body, is also suitable for hernia, lumbago, tinnitus, and neurosis (Chen and Tang, 1982; Wu, 1994; Hu et al.,2000).

10.5.2 Chronic Lung Disease

The chronic disease is characterized by an enduring illness that can be controlled but not completely cured. In the realm of herbal medicine, various *Habenaria* species have been employed to address chronic conditions. For instance, the tubers of *H. intermedia* and *H. plantaginea* were used in folk medicine to alleviate chronic diseases such as asthma. Furthermore, the flowers of *H. diphylla*, commonly referred to as "Jeevahi Purusharatna," were found to serve a similar purpose in addressing chronic ailments (Maridass et al., 2008; Singh and Duggal, 2009). *H. arietina* and *H. ciliolaris* were traditionally employed as remedies for tussis or coughing ailments. The whole plant of *H. pectinata* was utilized to treat discomforts like waist pain and persistent coughs arising from weakness. *H. dentata* and *H. petelotii* benefit the lungs and relieve coughs caused by tuberculosis (Da Cidian, 1986; Ou et al., 2003). Additionally, a decoction made from approximately 15 grams of *H. linguella* powder was employed to alleviate lung-related discomforts attributed to excess heat (Wu, 1994; Hu et al., 2000).

10.5.3 Arthritis

Arthritis is the inflammation that occurs in the joints or articulation during a movement. Incorporating condiments with *H. arietina* tubers represents a beneficial approach for arthritis relief. *H. pectinata* offers a natural remedy for a spectrum of ailments, including arthritis, fever, phthisis, diathesis bleeding, and burning sensation (Singh and Duggal, 2009).

10.5.4 Ulcer

Habenaria marginata, popularly known as "Humari" in India and Bangladesh, was used to treat malignant ulcers by the tribes in the Niyamgiri Hills of Orissa. To administer this treatment, tubers were boiled and the resulting decoction was given with a teaspoon of honey. This practice was followed for a duration of two weeks, and its efficacy has been documented in studies by Dash et al. (2008) and Hossain (2009). Chinese practitioners used *H. rhodocheila* paste for finger ulcers, and decoction prepared from this orchid was applied to facilitate the healing of traumatic injuries and reduce swellings (Wu, 1994; Hu et al., 2000).

10.5.5 Kidney-related Disease

Chinese Herbal Medicine has reported various uses of *Habenaria* in regulating kidney-related issues. Among these species, *H. arientina* is employed in the treatment of kidney inflammation, while *H. stenopetala* is known to enhance kidney and sexual functions (Hu et al., 2000). Furthermore, *H. limprichtii* serves as a valuable nutrition for women and aids in the improvement of renal functions.

The stems of *H. ciliolaris* and *H. dentata* are utilized for detoxification and ameliorating the gradual loss of kidney function. Additionally, the roots of *H. davidii* and *H. diplonema* play a role in regulating menstruation and also protect the kidneys (Wu, 1994). *H. aitchisonii* finds its application in the treatment of hematuria, with the roots being used to address nephritis. Lastly, *H. delavayi* tubers and *H. pectinate* are suitable for waist pain, nephritis, and lumbago (Chen and Tang, 1982; Da Cidian, 1986; Hu et al., 2000).

10.5.6 Venereal Disease and Vaginal Haemorrhage

H. floribunda is commonly called "clavocochinillo" (nail piglet) in Veracruz, Mexico, where the leaf infusion is taken orally for vaginal haemorrhage (Santos, 1988). Folklore from North Andaman stated that the tuber extracts of *H. andamanica*, locally called Bharga muli, are taken orally with pepper powder to treat venereal disease (Prasad et al., 2008).

10.5.7 Spermatorrhoea

The condition characterized by involuntary loss of semen is known as spermatorrhoea. Leaves and roots of *H. arietina* and *H. commelinifolia* (roots) act like aphrodisiacs, which are utilized in the treatment of spermatorrhoea (Rao and Sridhar, 2007).

In Chinese herbal medicine, *H. ciliolaris* and *H. dentata* are prescribed for the management of impotence (Chuakul, 2002). Furthermore, *H. grandifloriformis* tubers are powdered and mixed with milk to increase sperm count and for galactagogue. Folk medications suggest *H. stenopetala* and *H. petelotii* as a remedy for erectile dysfunction (Hu et al., 2000). These natural therapies highlight the richness of traditional medicine in addressing conditions mainly related to male reproductive health.

10.5.8 Wasting Disease

Ayurvedic physicians use *H. roxburghii, H. plantaginea*, and *H. furcifera* tubers for the treatment of wasting disease, fever, fainting, and haemorrhage (Yoganarasimhan and Chelladurai, 2000). *H. plataginea* tubers are also used to treat helminthiasis, paralysis, and tuberculosis. The Bangladeshis formulated a tablet with black pepper, garlic, and *H. plantaginea* tubers paste to alleviate continual chest pain and stomach ache (Hossain, 2009).

10.5.9 Hepatoprotective Activity

Hepatoprotective agents are substances capable of protecting the liver from the damage caused by hepatotoxins. The people from Yunnan, a southwestern province of China, utilize *H. diplonema* in their traditional folk medicine practices to improve liver function. It is commonly called "Shuangxianerye Lan," which means two-leafed orchid. *H. diplonema* roots act as a natural remedy to regulate the menstrual cycle in women (Wu, 1994).

10.5.10 Antivenom

H. ciliolaris H. plantaginea, and *H. pantlingiana* are applied for snake bites (Reddy et al., 2005; Maridass et al., 2008; Singh and Duggal, 2009). In India, ethnic communities have used the leaves and tubers of *H. pectinata* for the treatment of snake bites. Konda Reddis in Andra Pradesh relies on a combination of pepper, garlic, and *H. roxburghii* tubers for a venomous bite (Reddy et al., 2005). Similarly, the Chenchus tribe employs *H. furcifera* for a snake attack. *H. hollandiana* paste is employed for scorpion venom and maggot boils in Bangladesh (Hossain, 2009). These home remedies illustrate the diverse ways in which different *Habenaria* species contribute to the management of snake and scorpion-related health issues in various regions.

TABLE 10.1
Chemical Constituents of the Genus *Habenaria*

Active Chemical Constituents	Orchid Name	Biological Properties	Reference
Sinapic acid	*H. intermedia*	Anticancer, antimutagenic, antioxidant, anti-inflammatory, antiglycemic, antibacterial, and neuroprotective	(Chen, 2016)
Scopoletin	*H. intermedia*	Antioxidant, antimicrobial, anti-inflammatory, anti-cariogenic, immunomodulator, anti-metabolic disorder, blood purifier, and neuroprotective activity	(Antika et al., 2022; Habbu et al., 2012)
2-(2-hydroxy-6-methoxyphenethyl)-5-methoxybenzene-1,3-diol	*H. pectinate*	Antimicrobial activity	(Kaur, 2022)
Loroglossol	*H. petalodes*	Antioxidant, antimicrobic, anticancer, and immunostimulatory	(Bazzicalupo et al., 2023)
Isorhamnetin 3,7-di-O-β-D-gluco pyranoside	*H. petalodes*	Antioxidant activity	(Yokozawa et al., 2002; Cota et al., 2008)
Habenariol	*H. repens*	Feeding deterrent activity	(Wilson et al., 1999)
Gastrodigenin	*H. dentata*	Anti-hepatotoxic activity	(Li et al., 2019)
Hecogenin	*H. marginata*	Antiproliferative, antioxidant, anti-inflammatory, and neuroprotective properties	(Wu et al., 2021)

10.5.11 Leucoderma

The appearance of white patches or loss of skin pigmentation in skin that is not self-generated but rather occurs accidentally, resulting in depigmented areas on the skin, is known as leucoderma. *H. longicorniculata,* often referred to as Devasunda in India, is used to control pain, swelling, and infected sites of leukoderma. A decoction is prepared with 10 g of turmeric powder and 10 g of tuber extract of *H. longicorniculata* and applied externally to cure leukoderma (Rao and Sridhar, 2007; Dash et al., 2008).

10.5.12 Other Beneficial Properties

In addition to its therapeutic values, *Habenaria* exhibits other noteworthy properties such as anticancer, antioxidant, anti-inflammatory, antiglycemic, antibacterial, antimutagenic, and neuroprotective activities (Table 10.1). Recent research by Satish and Mallya (2019) discovered that the tuber of *H. longicorniculata* exhibited inhibitory activity against hydrogen peroxide. The larvicidal activity (mosquito) of *H. plantaginea* leaf extract was successful when combined with fabricated nanoparticles (Aarthi et al., 2018). Analgesic and anti-inflammatory properties were observed in *H. digitata* (Mahnashi et al., 2021). Furthermore, Ayurveda describes *H. intermedia* as an effective anti-stress and adaptogen agent (Habbu et al., 2012).

10.6 CONCLUSION AND FUTURE REMARKS

Habenaria, a widely distributed terrestrial orchid, has been examined globally with a focus on phytochemicals and pharmacological activities. The reviews emphasize the overall potential use of this orchid in disease management and treatments (Table 10.2), justifying its inclusion in traditional

TABLE 10.2
Medicinal Uses of the Genus *Habenaria*

S. No.	Species Name	Plant Parts Used	Therapeutic Uses	Reference
1.	*H. aitchisonii*. H.G. Reich.	Roots	In Chinese Herbal Medicine (CHM), it is used for healing nephritis and hematuria.	(Wu, 1994; Hu et al., 2000)
2.	*H. andamanica* Hook. f.	Tuber	Tuber extract is taken orally with pepper to treat venereal diseases.	(Prasad et al., 2008)
3.	*H. arietina* Hook. f.	Whole	Generally used as a blood purifier and also to treat kidney inflammation.	(Puri, 1970; Jalal et al., 2008)
		Rhizome	Aphrodisiac property.	(Puri, 1970; Jalal et al., 2008)
		Tuber	Emollient and appetiser. Tubers are used to relieve thirst, fever, coughs, asthma, worms, anorexia, emaciation, haematemesis, general debility, skin diseases, leprosy, insanity, and cataplexy. In Nepal, people use tubers to promote vitality and consume as a vegetable.	(Manandhar, 2002; Singh and Duggal, 2009)
4.	*H. ciliolaris* Kraenzl.	Herb	Used for ailments like stomach ache, tuberculous cough, hernia, urinary problems, kidney infections, nocturnal emissions, impotence, leucorrhoea, gonorrhea, and snake bites.	(Ou et al., 2003; Hu et al., 2000)
		Stem	Used as a detoxifier and reduces heat. Strengthens weak kidneys.	
5.	*H. cirrhata* (Lindl.) Rchb.f.	--	The Fulani tribe of West Africa calls its Balasa, which is used for bracing.	(Assédé et al., 2017)
6.	*H. commelinifolia* (Roxb.) Wall ex Lindl.	Whole	Consumed as a vegetable. It is used to cure blebs on the palm and is alleged to be a blood purifier.	(Kumar, 2003; Rao and Sridhar, 2007)
		Root	Treated for spermatorrhoea.	
7.	*H. crinifera* Lindl.	Tuber	Tribes of Kudremukh National Park in Karnataka use the tubers for headache.	(Rao and Sridhar, 2007)
8.	*H. davidii* Franch.	Whole	A decoction is made with the orchid for lymph node swellings and hernia.	(Wu, 1994; Hu et al., 2000)
		Root	In CHM, roots are used to reduce to reduce kidney swelling.	
9.	*H. delavayi* Finet	Whole plant	The tonic is used to strengthen the body and treat lumbago, dizziness, tinnitus, hernia, and neurosis.	(Chen and Tang, 1982; Wu, 1994)
		Tuber	Tubers are helpful in treating renal insufficiency, waist pain, and nephritis.	(Da Cidian, 1986; Hu et al., 2000)
10.	*H. dentata* (Sw.) Schltr.	Stem	Used to treat orchitis, dysuria, carbuncles, impotence, and coughs caused by tuberculosis.	(Chuakul, 2002)
		Tuber	Tubers are utilized by Thai herbalists to cure abscesses and bodily discomfort.	

(Continued)

TABLE 10.2 *(Continued)*
Medicinal Uses of the Genus *Habenaria*

S. No.	Species Name	Plant Parts Used	Therapeutic Uses	Reference
11.	*H. diphylla* (Nimmo) Dalzell	Whole plant	Home medicine for insect bites in Thailand.	(Chuakul, 2002)
		Flower	The tribals from the Western Ghats region of India use these Jeevahi Purusharatna flowers for asthma problems.	
12.	*H. diplonema* Schltr.	Root	It aids in regulating liver, kidney, and menstruation.	(Wu, 1994)
13.	*H. dives* Rchb. f.	Tuber	The lethal effects are induced by the use of dried tuber powder in combination with other ingredients.	(Hulme, 1954)
14.	*H. dregeana* Lindl	Leaf	Leaf infusions are applied for lightning.	(Hulme, 1954)
15.	*H. edgeworthii* Hook.f. ex Collett.	Leaves and roots	It provides a cooling effect and is spermophytic.	(Singh and Duggal, 2009)
16.	*H. epipactidea* Rchb.f.	Whole plant	The dried powder mixed with sheep fat is used to cure incisions cut into the foreheads, wrists, and ankles of every member of the homestead as protection from storms.	(Watt and Breyer-Brandwijk, 1962)
		Root	Root extract promotes vomiting (emetic) sensation.	(Hutchings et al., 1996)
17.	*H. floribunda* Lindl.	Leaf	The leaf infusion is used to relieve the pain of heavy menstruation.	(Santos, 1988)
18.	*H. fordii* Rolfe	Root	The roots help in treating indigestion in children	(Wu, 1994)
19.	*H. furcifera* Lindl.	Tuber	Curative tubers are used in Ayurveda to heal wasting disease, fever, blood disorder, haemorrhage, and fainting. The Chenchus tribe used the herbal paste as an antidote against snake poison.	(Yoganarasimhan and Chelladurai, 2000; Reddy et al., 2005)
20.	*H. grandifloriformis* Blatt. & McCann	Tuber	Tribals of the Melghat region extensively exploit this tuber during food scarcity as it is believed to control hunger. The fresh tubers are boiled and eaten as food. Tuber powder is mixed with milk to increase sperm count and galactagogue.	(Bhogaonkar and Devankar, 2007)
21.	*H. hollandiana* Santapau	Plant	The fresh paste is used by Kondareddies and Valmikis of Andra Pradesh to treat scorpion stings. Similarly, in Bangladesh, they use it for maggot infections.	(Akarsh, 2004; Hossain, 2009)
		Tuber	The Pawra tribe from Maharashtra practice some magical arts with these tubers.	(Jagtap et al., 2008)
22.	*H. intermedia* D. Don	Tuber	Tubers act as aphrodisiacs, rejuvenators, and depuratives and are used for several chronic diseases, such as asthma, leprosy, skin disease, and memory loss.	(Arora et al., 2023)

(Continued)

TABLE 10.2 *(Continued)*
Medicinal Uses of the Genus *Habenaria*

S. No.	Species Name	Plant Parts Used	Therapeutic Uses	Reference
23.	*H. limprichtii* Schltr.	Whole plant	In China, people considered this orchid to enhance "feminine nourishment" and to heal nephritis and renal function.	(Wu, 1994)
24.	*H. linguella* Lindl.	Whole plant	Herbal decoction is taken orally for heaty lungs.	(Hu et al., 2000)
25.	*H. longicorniculata* J Graham	Whole plant	The plant extract is mixed with turmeric and applied externally to cure leukoderma.	(Subramani and Goraya, 2003; Rao and Sridhar, 2007; Dash et al., 2008)
		Tuber	Reduce scrotal enlargement.	
26.	*H. macrandra* Lindl.	Root	Boiled roots were used as purgative in East Africa.	(Assédé et al., 2017)
27.	*H. marginata* Colebr.	Tuber	The tribals from Niyamgiri Hills of Orissa and Bangladesh use this tuber to treat malignant ulcers. It is also used as a vegetable at the Sanjay National Park in Madhya Pradesh.	(Kumar, 2003; Dash et al., 2008; Hossain, 2009)
28.	*H. multipartita* Bl. Ex Kraenzl	Tuber	Tubers are used as an edible product.	(Heyne, 1987)
29.	*H. pantlingiana* Kraenzl.	–	Antivenom, nociception, inflammation, and pyrexia.	(Hegde, 1984; Mahnashi et al., 2021)
30.	*H. pectinata* D. Don	Whole plant	Helps to reduce pain at the waist and lessen coughs arising from weakness.	(Wu, 1994; Hu et al., 2000; Singh and Duggal, 2009)
		Leaf	Antivenom property.	
		Tuber	Tuber extracts are used as a natural remedy for healing arthritis, diathesis bleeding, fever, burning sensation, and phthisis.	
31.	*H. petelotii* Gagnep.	Whole plant	CHM reports the use of *H. petelotii* for curative effect against hernia, nocturnal bed-wetting in children, erectile dysfunction, and external injuries with bleeding.	(Wu, 1994; Hu et al., 2000)
32.	*H. plantaginea* Lindl.	Tuber	In India, tubers are used to treat helminthiasis, cough, asthma, wasting disease, fever, haemorrhages, fainting tuberculosis, and paralysis. In Bangladesh, tubers are consumed to have a curative effect on chest pain and stomach ache.	(Rao and Henry, 1996; Yoganarasimhan and Chelladurai, 2000; Maridass et al., 2008; Hossain, 2009; Singh and Duggal, 2009; Hossain, 2011)
33.	*H. rariflora* A. Rich.	Tuber	Tubers may have a curative effect on healing wasting disease, fever, blood disorders, fainting, and haemorrhage.	(Yoganarasimhan and Chelladurai, 2000)
34.	*H. rumphii* (Brongn.) Lindl.	Tuber	Delicious jams or candy are made with these tubers.	(Rafai, 1975)
35.	*H. rhodocheila* Hance	Whole plant	Tribals from China used the plant extract to cure finger ulcers. The decoction is made with 9 g of herbal powder to relieve pain, swelling, and traumatic injuries.	(Wu, 1994; Hu et al., 2000)

(Continued)

TABLE 10.2 *(Continued)*
Medicinal Uses of the Genus *Habenaria*

S. No.	Species Name	Plant Parts Used	Therapeutic Uses	Reference
36.	*H. roxburghii*, Nicolson	Tuber	The Dravidian ethno-linguistic group use *H. roxburghii* similar to *H. rariflora* and *H. furcifera*. The Konda Reddis in Andra Pradesh identified the antivenom property of this orchid and supply it with pepper and garlic.	(Yoganarasimhan and Chelladurai, 2000; Reddy et al., 2005; Rao et al., 2011)
37.	*H. schimperiana* Hochst. ex A.Rich.	--	Fulani women used this orchid for eyesight improvement.	(Assédé et al., 2017)
38.	*H. stenopetala* Lindl.	Plant	The plant is used by Chinese herbalists to treat erectile dysfunction, kidney dysfunction, sexual dysfunction, and hernias.	(Hu et al., 2000)
39.	*H. walleri* Rchb.f.	Root	The roots were administered for stomach disorders by East Africans.	(Stewart and Campbell, 1996)

systems of medicine. So far, only a few areas of research, such as vegetative anatomy, floral morphology, and *in vitro* propagation of some species, have been reported so far. The taxonomical studies have been limited to a few countries, such as India, Brazil, Timor-Leste, and Thailand. Only the economic importance of *H. edgeworthii* and *H. intermedia* has been extensively studied, and many potentially bioactive species are overlooked. Traditional knowledge of orchids has provided valuable insights into the plant's effectiveness against various diseases, but further in-depth research is essential to substantiate its pharmacological trials under in-vivo conditions. To identify the active bioactive compounds responsible for the medicinal importance of *Habenaria* species accurately, modern research activities are necessary. Such detailed studies hold the potential to accelerate the drug discovery process more precisely, offering new avenues for medical treatment and management.

REFERENCES

Aarthi, C., Govindarajan, M., Rajaraman, P., Alharbi, N. S., Kadaikunnan, S., Khaled, J. M., Mothana, R. A., Siddiqui, N. A., & Benelli, G. (2018). Eco-friendly and cost-effective Ag nanocrystals fabricated using the leaf extract of *Habenaria plantaginea:* Toxicity on six mosquito vectors and four non-target species. *Environmental Science and Pollution Research, 25*, 10317–10327.

Akarsh, M. (2004). Indian medicinal plants. *Newsletter of Envis Node, 1*(2), 49.

Antika, L. D., Tasfiyati, A. N., Hikmat, H., & Septama, A. W. (2022). Scopoletin: A review of its source, biosynthesis, methods of extraction, and pharmacological activities. *Zeitschrift für Naturforschung C, 77*(7–8), 303–316.

Arora, M., Arora, K., & Kaur, R. (2023). Pharmacognostic, physicochemical, phytochemical, nutraceutical evaluation and in vitro antioxidant potency of *Habenaria intermedia* (D. Don) Szlach-A rare orchid. *South African Journal of Botany, 152*, 278–287.

Assédé, E. S. P., Djagoun, C. A. M. S., Azihou, A. F., Kouton, M. D., Gogan, Y. S. C., Geldenhuys, C. J., Chirwa, P. W., & Sinsin, B. A. (2017). Folk perceptions and patterns of use of orchid species in Benin, West Africa. *Flora Et Vegetatio Sudano-Sambesica, 20*, 26–36.

Bazzicalupo, M., Calevo, J., Smeriglio, A., & Cornara, L. (2023). Traditional, therapeutic uses and phytochemistry of terrestrial european orchids and implications for conservation. *Plants, 12*(2), 257.

Bhogaonkar, P. Y., & Devankar, V. D. (2007). Pharmacognostic studies on Padmacarini. *Aryavaidyan, XX*(2), 74–79.

Bungtongdee, N., Sopalun, K., Laosripaiboon, W., & Iamtham, S. (2019). The chemical composition, antifungal, antioxidant and antimutagenicity properties of bioactive compounds from fungal endophytes associated with Thai orchids. *Journal of Phytopathology*, *167*(1), 56–64.

Chen, C. (2016). Sinapic acid and its derivatives as medicine in oxidative stress-induced diseases and aging. In: *Oxidative Medicine and Cellular Longevity*. London: Hindaw.

Chen, S. C., & Tang, T. (1982). A general review of the orchid flora of China. In: Arditti, J. (Ed.), *Orchid Biology: Reviews and Perspectives*, Vol. II. Ithaca, NY: Cornell University Press, pp. 39–87.

Chen, X. H., Tan, S. L., Liang, Y. L., Huang, L., Xiao, H. W., Luo, H. L., Xiong, D. J., Yang, B. Y., & Ren, Z. X. (2021). The pollination of *Habenaria rhodocheila* (Orchidaceae) in South China: When butterflies take sides. *Ecology and Evolution*, *11*(6), 2849–2861.

Chen, X. Q., & Cribb, P. J. (2009). *Habenaria* Willdenow. In: Chen, X. Q., Liu, Z. J., Zhu, G. H., Lang, K. Y., Ji, Z. H., Luo, Y. B., Jin, X. H., Cribb, P. J., Wood, J. J., Gale, S. W., & Ormerod, P. (Eds.), *Flora of China – Orchidaceae*. Beijing: Science Press.

Chuakul, W. (2002). Ethnomedical uses of Thai orchidaceous plants. *Mahidol University Journal of Pharmaceutical Sciences*, *29*(3–4), 41–45.

Cota, B. B., Magalhães, A., Pimenta, A., Siqueira, E. P., Alves, T., & Zani, C. L. (2008). Chemical constituents of *Habenaria petalodes* Lindl. (Orchidaceae). *Journal of the Brazilian Chemical Society*, *19*, 1098–1104.

Da Cidian, Z. (1986). *Edited by Jiangsu New Medical College*. Shanghai: Science and Technology Press.

Dash, P. K., Sahoo, S., & Bal, S. (2008). Ethnobotanical studies on orchids of Niyamgiri hill ranges, Orissa, India. *Ethnobotanical Leaflets*, *2008*(1), 9.

Davis, R. S., & Steiner, M. L. (1982). *Philippine Orchids: A Detailed Treatment of Some One Hundred Native Species*. M & L Licudine Enterprises.

Eldin, A. B., Ezzat, M., Afifi, M., Sabry, O., & Caprioli, G. (2023). Herbal medicine: The magic way crouching microbial resistance. *Natural Product Research*, *37*, 4280–4289.

Girija, P. L. T., & Sivan, N. (2022). Ayurvedic treatment of COVID-19: A case report. *Journal of Ayurveda and Integrative Medicine*, *13*(1), 100329.

Gupta, V., Kaur, G., Maithani, M., Bansal, R., & Bansal, P. (2023). "Ashtawarga": The wonder medicinal herbs of nature. In: *Herbs, Spices and Their Roles in Nutraceuticals and Functional Foods* (pp. 309–336). Cambridge, MA: Academic Press.

Habbu, P. V., Smita, D. M., Mahadevan, K. M., Shastry, R. A., & Biradar, S. M. (2012). Protective effect of *Habenaria intermedia* tubers against acute and chronic physical and psychological stress paradigms in rats. *Revista Brasileira de Farmacognosia*, *22*, 568–579.

Hegde, S. N. (1984). *Orchid of Arunachal Pradesh*. Arunachal Pradesh: Department of Forests, Government of Arunachal Pradesh.

Heyne, K. (1987). The useful plants of Indonesia. In *Badan Litbang Kehutanan*, Vols. I, III, and IV. Badan Penelitian dan Pengembangan Kehutanan (in Indonesian, translated from Dutch).

Hossain, M. M. (2009). Traditional therapeutic uses of some indigenous orchids of Bangladesh. *Medicinal and Aromatic Plant Science and Biotechnology*, *42*(1), 101–106.

Hossain, M. M. (2011). Therapeutic orchids: Traditional uses and recent advances—An overview. *Fitoterapia*, *82*(2), 102–140.

Hu, X. M., Zhang, W. K., Zhu, Q. Z., et al. (2000). *Zhonghua Bencao*, Vol. 8. Shanghai: Shanghai Science and Technology Publication.

Hulme, M. (1954). *Wild Flowers of Natal*. Pietermaritzburg: Schuter and Shooter.

Hutchings, A., Scott, A. H., Lewis, G., & Cunningham, A. B. (1996). *Zulu Medicinal Plants: An Inventory*. Pietermaritzburg: University of Natal Press.

Jagtap, S. D., Deokule, S. S., & Bhosle, S. V. (2008). Ethnobotanical uses of endemic and RET plants by Pawra tribe of Nandurbar district, Maharashtra. *Indian Journal of Traditional Knowledge*, *7*(2), 311–315.

Jalal, J. S., Kumar, P., & Pangtey, Y. P. S. (2008). Ethnomedicinal orchids of Uttarakhand, Western Himalaya. *Ethnobotanical Leaflets*, *2008*(1), 164.

Kanlayavattanakul, M., & Lourith, N. (2022). Orchid extracts and cosmetic benefits. In: *Orchids Phytochemistry, Biology and Horticulture: Fundamentals and Applications* (pp. 609–626). Cham: Springer International Publishing.

Kaur, C. (2022). Evaluation of the antimicrobial potential of methanolic extract and a dihydrostilbene isolated from tubers of *Habenaria pectinata*. *Research Journal of Pharmacy and Technology*, *15*(5), 2307–2312.

Kumar, C., Chandan, G., Kushwaha, M., Kumar, A., Kaur, S., Kumar, A., Yadav, G., Gairola, S., Vishwakarma, R. A., Satti, N. K., & Verma, M. K. (2023). Discovery of anti-NRLP3 inflammasome, immunomodulatory phytochemicals from the extract of *Habenaria intermedia* D. Don: An unexplored plant species. *ACS Omega*, *8*(34), 31112–31122.

Kumar, V. (2003). Wild edible plants of Surguja district of Chhattisgarh state, India. *Journal of Economic and Taxonomic Botany*, *27*(2), 272–282.
Li, K., Wu, F., Chen, M., Xiao, Z., Xu, Y., Xu, M., Liu, J., & Xu, D. (2023). Identification, biological function profiling and biosynthesis of secondary metabolites in medicinal orchids. *Metabolites*, *13*(7), 829.
Li, W., Sun, C., Deng, W., Liu, Y., Adu-Frimpong, M., Yu, J., & Xu, X. (2019). Pharmacokinetic of gastrodigenin rhamnopyranoside from Moringa seeds in rodents. *Fitoterapia*, *138*, 104348.
Lihua, T. (2015). *Application of Traditional Chinese Medicine Composition in Preparation of Anti-Tuberculin Medicine* (China Application No. CN104815175A) (https://patents.google.com/patent/CN104815175A/en).
Liu, H., Luo, Y. B., Heinen, J., Bhat, M., & Liu, Z. J. (2014). Eat your orchid and have it too: A potentially new conservation formula for Chinese epiphytic medicinal orchids. *Biodiversity and Conservation*, *23*, 1215–1228.
Mahnashi, M. H., Alyami, B. A., Alqahtani, Y. S., Jan, M. S., Rashid, U., Sadiq, A., & Alqarni, A. O. (2021). Phytochemical profiling of bioactive compounds, anti-inflammatory and analgesic potentials of *Habenaria digitata* Lindl.: Molecular docking based synergistic effect of the identified compounds. *Journal of Ethnopharmacology*, *273*, 113976.
Manandhar, N. P. (2002). *Plants and People of Nepal*. Portland, OR: Timber Press.
Maridass, M., Hussain, M. I., & Raju, G. (2008). Phytochemical survey of orchids in the Tirunelveli hills of South India. *Ethnobotanical Leaflets*, *2008*(1), 95.
Medhi, R. P., & Chakrabarti, S. (2009). Traditional knowledge of NE people on conservation of wild orchids. *Indian Journal of Traditional Knowledge*, *8*(1), 11–16.
Menzepoh, S. B. (2011). Edible orchids by the bagam tribe in Cameroon. *Biotechnologie, Agronomie, Société et Environnement*, *15*(4), 509–514.
Nurfadilah, S. (2020, March). Utilization of orchids of Wallacea region and implication for conservation. In: *IOP Conference Series: Earth and Environmental Science* (Vol. 473(1), p. 012063). Bristol: IOP Publishing.
Ou, J. C., Hsieh, W. C., Lin, I. H., Chang, Y. S., & Chen, I. S. (2003). *The Catalogue of Medicinal Plant Resources in Taiwan*. Taipei: Department of Health, Executive Yuan, p. 150.
Prasad, P. R. C., Reddy, C. S., Raza, S. H., & Dutt, C. B. S. (2008). Folklore medicinal plants of north Andaman Islands, India. *Fitoterapia*, *79*(6), 458–464.
Pridgeon, A. M., Cribb, P. J., Chase, M. W., & Rasmussen, F. (2001). *Genera Orchidacearum: Volume 2. Orchidoideae (Part 1)*. Genera Orchidacearum.
Puri, H. S. (1970). Indian medicinal plants used in elixirs and tonics. *Quarterly Journal of Crude Drug Research*, *10*(2), 1555–1566.
Qiu, L., Jacquemyn, H., Burgess, K. S., Zhang, L. G., Zhou, Y. D., Yang, B. Y., & Tan, S. L. (2023). Contrasting range changes of terrestrial orchids under future climate change in China. *Science of the Total Environment*, *895*, 165128.
Rafai, M. A. (1975). *Extraordinary Uses of Orchids in Indonesia*. Report of the First Asean Orchid Congress. Bangkok: Ministry of Agriculture, Ministry of Foreign Affairs and Kasetsart University.
Rao, J. K., Suneetha, J., Reddi, T. S., & Kumar, O. A. (2011). Ethnomedicine of the Gadabas, a primitive tribe of Visakhapatnam district, Andhra Pradesh. *International Multidisciplinary Research Journal*, *1*(2).
Rao, N. R., & Henry, A. N. (1996). *Ethnobotany of Eastern Ghats in Andhra Pradesh, India*. Botanical Survey of India. Calcutta: Govt. Press of India.
Rao, T. A., & Sridhar, S. (2007). *Wild Orchids in Karnataka: A Pictorial Compendium*. Bangalore: Institute of Natural Resources Conservation Education, Research and Training.
Rawat, S., Andola, H., Giri, L., Dhyani, P., Jugran, A., Bhatt, I. D., & Rawal, R. S. (2014). Assessment of nutritional and antioxidant potential of selected vitality strengthening Himalayan medicinal plants. *International Journal of Food Properties*, *17*(3), 703–712.
Reddy, K. N., Reddy, C. S., & Raju, V. S. (2005). Ethno-orchidology of orchids of Eastern Ghats of Andhra Pradesh. *EPRTI Newsletter*, *11*(3), 5–9.
Santos, R. M. A. (1988). *Ethnobotany: Medicinal Plants of Zoque-Popolucas of Ejidos of San Fernando, Santa Marta, Ocotal Grande and Ocotal Chico, Soteapan, Municipality, Veracruz*. Bachelor Thesis, Biology Faculty, Veracruz University, Xalapa, Veracruz (in Spanish).
Satish, B. N., & Mallya, S. V. (2019). In-vitro antioxidant activity assay of *Habenaria longicorniculata* J. Graham wild medicinal tubers. *Journal of Drug Delivery and Therapeutics*, *9*(3S), 342–344.
Singh, A., & Duggal, S. (2009). Medicinal orchids - An overview. *Ethnobotanical Leaflets*, *2009*(3), 3.
Singh, S. K., Agrawala, D. K., Jalal, J. S., Dash, S. S., Mao, A. A., & Singh, P. (2019). *Orchids of India: A Pictorial Guide*. Botanical Survey of India, Ministry of Environment, Forest and Climate Change.

Stewart, J., & Campbell, B. (1996). *Orchids of Kenya*. Winchester: St Pauls Bibliographies, p. 176.

Subramani, S. P., & Goraya, G. S. (2003). Some folklore medicinal plants of Kolli Hills: Record of a Natti Vaidyas Sammelan. *Journal of Economic and Taxonomic Botany*, *27*(3), 665–678.

Watt, J. M., & Breyer-Brandwijk, M. G. (1962). The medicinal and poisonous plants of Southern and Eastern Africa being an account of their medicinal and other uses, chemical composition, pharmacological effects and toxicology in man and animal. In: *The Medicinal and Poisonous Plants of Southern and Eastern Africa Being an Account of their Medicinal and Other Uses, Chemical Composition, Pharmacological Effects and Toxicology in Man and Animal* (2nd ed.). London: E. & S. Livingstone.

Wilson, D. M., Fenical, W., Hay, M., Lindquist, N., & Bolser, R. (1999). Habenariol, a freshwater feeding deterrent from the aquatic orchid *Habenaria repens* (Orchidaceae). *Phytochemistry*, *50*(8), 1333–1336.

Wu, S., Pang, Y., He, Y., Zhang, X., Peng, L., Guo, J., & Zeng, J. (2021). A comprehensive review of natural products against atopic dermatitis: Flavonoids, alkaloids, terpenes, glycosides and other compounds. *Biomedicine & Pharmacotherapy*, *140*, 111741.

Wu, X. R. (1994). *A Concise Edition of Medicinal Plants in China*. Guangdong, China: Guangdong Higher Education Publication House (In Chinese).

Yoganarasimhan, S. N., & Chelladurai, V. (2000). *Medicinal Plants of India: Tamil Nadu* (Vol. 2). Bangalore: Regional Research Institute.

Yokozawa, T., Kim, H. Y., Cho, E. J., Choi, J. S., & Chung, H. Y. (2002). Antioxidant effects of isorhamnetin 3, 7-di-O-β-D-glucopyranoside isolated from mustard leaf (*Brassica juncea*) in rats with streptozotocin-induced diabetes. *Journal of Agricultural and Food Chemistry*, *50*(19), 5490–5495.

11 Curative and Restorative Properties of *Luisia* Orchids

Pawan Singh Rana

11.1 INTRODUCTION

Orchids are one of the most diverse and captivating plant families, boasting over 25,000 species worldwide. Their exquisite beauty and enchanting fragrance have long captivated the human imagination. However, beyond their aesthetic appeal, these delicate and intricate flowers hold a fascinating secret—their potential as a source of valuable medicinal compounds. Orchids are not only cherished for their ornamental value but also for the unique bioactive compounds present within them that have been used in traditional medicine for centuries. From the vibrant petals of orchids to the hidden chemical compounds within, these remarkable plants have much to offer in the field of healthcare and well-being.

Among the ancient civilizations, Ayurveda and Traditional Chinese Medicine (TCM) contain well-elaborated descriptions of the curative uses of orchids. Half of the ashtavarga plants (a group of eight important medicinal plants) described in Ayurveda are orchids viz. Jivaka (*Malaxis mucifera*), Rishbhaka (*Malaxis accuminate*), Riddhi (*Habernaria intermedia*) and Vriddhi (*Habernaria edgeworthii*). Phytochemical compounds like alkaloids, anthocyanins, stilbenoids, and triterpenoids are found present in them. Compounds specific to orchids, such as Orchinol, cypripedin, jibantine, loroglossin, nidemine, and hircinol, etc., are also isolated from orchids (Singh and Duggal, 2009).

Among this vast and beautiful family of flowering plants, *Luisia* orchids stand out as a distinct and intriguing genus. In this chapter, we will delve into the world of *Luisia* orchids, exploring their characteristics, habitat, cultivation, and unique features that make them a noteworthy group of orchids. Join us on this journey as we uncover the fascinating relationship between orchids and medicine, shedding light on their rich history and the promising avenues for future research and development in the realm of natural remedies and pharmaceuticals.

11.2 TAXONOMY AND CLASSIFICATION

Luisia was named after a 19th-century Spanish botanist, Don Luis de Torres; these orchids belong to the family Orchidaceae, which is renowned for its exceptional diversity. The genus encompasses approximately 40 known species (Ma et al., 2023). Their distribution is throughout Southeast Asia, India, Sri Lanka, and some parts of Australia with two endemic species; one is distributed along the northern region of Northern territory and the other in the northeastern areas of Queensland (Plants World Online). These are epiphytic in nature, usually erect and straggly. The leafy stem branches close at the base with a fibrous, flattish, thick, long root. The narrow and thick leaves are usually arranged in loose spirals, moderately spaced, forming two ranks. The short raceme inflorescences are with a club-shaped and thick main stalk. Flowers produced sporadically last for a few days only (ANBG) (Table 11.1).

11.3 DISTINCTIVE CHARACTERISTICS

Luisia orchids exhibit a range of unique features that set them apart from other orchid genera:

1. **Inflorescence**: *Luisia* orchids produce inflorescences that bear multiple, small, and colorful flowers. The inflorescences can be quite striking, often forming a pendant or raceme with many blooms.

DOI: 10.1201/b23375-11

TABLE 11.1
Botanical Features of *Luisia* Orchids

Characteristics	Description
Inflorescence	Multiple, small, colorful flowers; pendant or raceme arrangement
Flower structure	Small, various colors, intricate lip, or labellum
Growth habit	Monopodial growth, upward from a single stem
Adaptation	Thrives in high light conditions, well-suited to tropical regions
Habitat and distribution	Found in lowland and montane rainforests, coastal areas

2. **Flower Structure**: The individual flowers of *Luisia* orchids are typically small, ranging from one to two centimeters in size. They come in various colors, including shades of purple, pink, white, and yellow. The lip or labellum, a modified petal, often displays intricate patterns and structures, adding to their visual appeal.
3. **Growth Habit**: These orchids usually have a monopodial growth habit, meaning they grow upward from a single stem, which can become quite tall over time.
4. **Adaptation**: *Luisia* orchids are adapted to growing in high-light conditions and are well suited to bright, indirect light. Their unique adaptations allow them to thrive in tropical and subtropical regions.
5. **Habitat and Distribution**: *Luisia* orchids can be found in a variety of habitats, including lowland and montane rainforests, as well as coastal areas. They are often spotted growing on the trunks and branches of trees or on rocks. In their native regions, they can be epiphytic, lithophytic, or even terrestrial, depending on the species and specific environmental conditions.

11.4 BOTANICAL DESCRIPTION

Straggly epiphytic herbs are perennial and evergreen in nature. Usually adherent highly branched elongated thick and flattish roots. The short to long stems get leafy at the apex and relatively thicker. Leaves can last for several seasons, spaced moderately, sessile, thick, and relatively longer. Pseudo bulbs are absent in *Luisia*. A semi-tubular and truncate peduncle, either the same or shorter than the rhachis, having imbricate bracts, is present. Non-persistent and scarious floral bracts partly sheath the pedicel at the base. The rhachis is straight, thick, and clavate. The ovaries are short and curved and deeply merge with the short pedicel. Greenish Flowers with a red labellum are produced sporadically and resupinate. The thin textured and incurved perianth is relatively segmented. The lateral sepals remain free and are broader than the dorsal sepals flank the labellum. The dorsal sepals are also free but narrower than the laterals. Petals are longer and slenderer than sepals remain free. The labellum is highly different in shape and size to petals and sepals rigidly sticking to the base of the column. Lamina of the labellum is bifurcated to epichile and hypochile; epichile is broader than hypochile, complete, and plump, while hypochile is concave with insignificant lateral lobes. Spur is absent. The callus with a groove or median plate is present, which is sometimes vague. The fleshy, straight, and short column extends to the apex from the ovary and lacks style and free filament, while nectar is unknown. Wings in the column are either vestigial or absent (rarely vestigial). The persistent and smooth anther with a decurved and short rostrum is two-celled, incumbent, and terminal. Pollinarium with two stalked, waxy, orange, and sessile pollinia are present. Viscidium is small, and a short, relatively broad stipe is present. Stigma is complete and concave. Erect, glabrous, dehiscent, and elongated capsules are present. Numerous seeds are light, winged, and colored. Peduncle and pedicel are not elongated in fruit and flower, respectively (Balachandar et al., 2019).

11.5 BIOLOGY

1. **Flowering:** The native species of *Luisia* flower sporadically between spring and autumn.
2. **Pollination:** The flowers last for several days only, and native bees probably pollinate them.
3. **Seasonal Growth:** *Luisia* plants grow in summer and spring, only going relatively dormant for the rest of the year.
4. **Reproduction:** *Luisia* reproduces through seeds only, and after pollination, the dispersal occurs within 10–12 months, and the capsule develops in an upright position only.
5. **Hybrids:** Natural hybrids involving the native species of *Luisia* are unknown.

11.6 MEDICINAL PROPERTIES

There is limited scientific information available regarding the medicinal properties of *Luisia* orchids. *Luisia* orchids, like many other orchid species, are primarily cultivated for their ornamental value due to their attractive flowers, while their use in traditional medical practices and as a source of therapeutic compounds is not that common.

Orchids, in general, are used in some traditional medicine systems in different parts of the world for various purposes (Gutiérrez, 2010), but these uses are often based on local folklore and anecdotal evidence rather than scientific research. Some potential traditional uses of orchids, which may or may not apply to *Luisia* orchids, include:

1. **Aphrodisiac:** Orchids have been associated with aphrodisiac properties in some cultures, but there is no scientific evidence to support these claims.
2. **Skin Care:** In some traditional practices, orchids are used in skincare products, but their effectiveness for this purpose is largely unproven.
3. **Traditional Medicine:** In traditional Chinese medicine, Ayurveda, and other traditional healing systems, orchids have utility for various purposes, including as tonics and for promoting general well-being. However, specific uses and properties can vary widely (Rahman and Huda, 2021).

It is essential to exercise caution when considering the medicinal properties of orchids, including *Luisia* orchids. Orchids are generally grown for their beauty and are not typically considered a source of therapeutic compounds or remedies. If you are looking for natural remedies or medicines, it is advisable to consult with a qualified healthcare professional who can provide evidence-based guidance and recommend safe and effective treatments.

Some important species of the genus *Luisia* include:

Luisia antennifera Blume, 1849.
Luisia brachystachys (Lindl.) Blume, 1849.
Luisia tenuifolia Blume, 1849.
Luisia teres (Thunb.) Blume, 1849.
Luisia trichorrhiza (Hook.) Blume, 1849.
Luisia volucris Lindl., 1853.
Luisia confusa Rchb.f. in W.G.Walpers, 1863.
Luisia psyche Rchb.f., 1863.
Luisia zollingeri Rchb.f. in W.G.Walpers, 1863.
Luisia macrotis Rchb.f., 1869.
Luisia microptera Rchb.f., 1870.
Luisia primulina C.S.P.Parish and Rchb.f., T 1874.
Luisia filiformis Hook.f., 1890.

Luisia tristis (G.Forst.) Hook.f., 1890.
Luisia amesiana Rolfe, 1893.
Luisia cantharis Rolfe, 1895.
Luisia hancockii Rolfe, 1896.
Luisia morsei Rolfe, J. Linn. Soc., Bot. 36: 33 1903.
Luisia boninensis Schltr., 1906.
Luisia foxworthii Ames, 1908.
Luisia taurina J.J.Sm., 1910.
Luisia celebica Schltr., 1911.
Luisia ramosii Ames, 1911.
Luisia javanica J.J.Sm., 1914.
Luisia megasepala Hayata, 1914.
Luisia unguiculata J.J.Sm., 1926.
Luisia macrantha Blatt. and McCann, 1932.
Luisia cordatilabia Ames and Quisumb., P1933 publ. 1934.
Luisia cordata Fukuy., 1934.
Luisia jonesii J.J.Sm., 1943.
Luisia recurva Seidenf., 1971.
Luisia secunda Seidenf., 1971.
Luisia thailandica Seidenf., 1971.
Luisia abrahamii Vatsala in A.Abraham and P.Vatsala, 1981.
Luisia pulniana Vatsala in A.Abraham and P.Vatsala, 1981.
Luisia longispica Z.H.Tsi and S.C.Chen, 1994.
Luisia magniflora Z.H.Tsi and S.C.Chen, 1994.
Luisia curtisii Seidenf., 1997.
Luisia appressifolia Aver., 2000.

Pharmacognostic works on some of these *Luisia* species show their immense curative and restorative potential (Table 11.2). One of the prominent species that shows therapeutic potential is *Luisia zeylanica*. It is a widely distributed species of the *Luisia* genus and has been studied for its curative properties by scientists over time. It has displayed ethnomedicinal value in treating chronic boils, fractures, tumors, abscesses, and burns (Hossain, 2009). Sohag et al. (2017) displayed the antioxidative properties and phytochemical constituents of *L. zeylanica*, through their studies. Significant inhibition of cancerous cells was observed in the MTT assay using methyl and ethyl acetate extracts of *L. zeylanica* besides its antifungal and antibacterial potential with these two extracts, respectively (Jhansi et al., 2023). Researchers in Andhra Pradesh (India) have also tested the anticancer potential of *L. zeylanica* (Khasim et al., 2019).

Luisia trichorrhiza is another species of the genus found to have its uses in diarrhea, muscle pain, and jaundice (Hossain, 2009). The root paste of *L trichorhiza* is prepared with ginger and turmeric for use in jaundice and muscle pain treatment (Rajeshbhai, 2022). Still, there is no record of pharmacognostic examination of this species.

Luisia volucris is another species that has been shown to have wound-healing property contains an important phytochemical named Luisiasin (De, 2022). It is also used in the treatment of fever, cough, and headaches. *Luisia tristis* is one of the important medicinal orchids containing a small amount of alkaloid compounds. Crushed plant paste is used to treat boils and tumors in some parts of Uttar Pradesh, India. In Sri Lanka, it is used as a medicinal oil to heal fractures. The juice made out of plant leaves is used to remove worms and treat chronic wounds in Nepal. The stem and its parts are crushed to paste and applied to fractured body parts as well as boils and tumors in almost every part of southeast Asia wherever it is found (Balachandar et al., 2019). The dried plant paste with turmeric and ginger is given to treat jaundice in ten days. Further, root extracts of this plant

TABLE 11.2
Selected *Luisia* Orchid Species and Their Therapeutic Properties

Species	Traditional Uses
Luisia zeylanica	1. Treatment of chronic boils, fractures, tumors, abscesses, and burns (Hossain, 2009). Antioxidative properties (Sohag et al., 2017). 2. Inhibition of cancerous cells in MTT assay (Jhansi et al., 2023). 3. Antifungal and antibacterial potential (Jhansi et al., 2023). 4. Limited pharmacognostic studies on its therapeutic potential (Hossain, 2009; Jhansi et al., 2023).
Luisia trichorriza	1. Used for diarrhea, muscle pain, and jaundice (Hossain, 2009). 2. Root paste preparation with ginger and turmeric for jaundice and muscle pain treatment (Rajeshbhai, 2022). 3. Traditional uses are documented, but no pharmacognostic examination is available (Hossain, 2009; Rajeshbhai, 2022).
Luisia volucris	1. Wound healing properties. 2. Use in the treatment of fever, cough, and headaches (De, 2022). 3. Presence of Luisiasin, an important phytochemical with wound healing properties (De, 2022).
Luisia tritis	1. Used for treating boils, tumors, fractures, and chronic wounds. 2. Traditional use in Nepal for removing worms (Balachandar et al., 2019). Contains a small amount of alkaloid compounds. 3. Various traditional uses across South East Asia (Balachandar et al., 2019).
Luisia tenuifolia	1. Whole plant paste used for healing swellings. 2. Plant powder mixed with vinegar for gonorrhea, leucorrhea, and kidney diseases (Sethuraman and Ramachandran, 2022). 3. Traditional uses are documented, but limited pharmacognostic examination is available (Sethuraman and Ramachandran, 2022).
Luisia teres	1. Entire plant used to heal fractures and reduce swelling. 2. Ante-poison in cases of poisoning (Teoh, 2016). 3. Limited information on medicinal properties; traditional uses documented (Teoh, 2016).

are also used for diarrhea treatment and alleviating myalgia in cattle. The entire plant is made into paste with egg white, calcium, and turmeric as a plaster for settling fractured limbs in some parts of Orrisa, India (Saravanan et al., 2023).

Luisia brachystachys was found to have use as herbal medicine in rheumatic pain, boils, tumors, and as a nutrient. *Luisia curtisii* is known for its use as a diuretic in traditional medicine systems. *Luisia hancockii* is a species found mainly in China, used to reduce swelling and remove swelling and gas as well. *Luisia morsii,* in traditional Chinese medicine, is used as an energy booster and to treat vomiting. In Taiwan and some other southeast Asian countries, *L morsii* is traditionally used to treat malaria, hypertension, rheumatism, paralysis, and poison removal. It is also used to kill bugs in fields and living places (Teoh, 2016). *Luisia tenuifolia* has its presence in peninsular India, preferring an altitude of 1,500m. The whole plant paste is used to heal swellings. The plant powder is mixed with vinegar to treat gonorrhea, leucorrhea, and kidney diseases (Sethuraman and Ramachandran, 2022).

The entire plant of *Luisia teres is* used by Taiwanese herbalists to heal fractures and remove swellings. It is effective against malaria and carcinoma. In cases of poisoning, it can have effective uses as ante-poison (Teoh, 2016). *Luisia merchantha* is another known species that possesses medicinal ability. The whole plant is used as an antibacterial and in the treatment of fever, coughs, and headaches (Teoh, 2016).

Some common medicinal properties are further described herewith (Table 11.3).

TABLE 11.3
Medicinal Properties of *Luisia* Orchids

Medicinal Property	Species with medicinal value
Pain relief and anti-inflammatory effects	*Luisia brachystachys, Luisia teres*
Antibacterial and antimicrobial	*Luisia javanica, Luisia tristis*
Antioxidant and neuroprotective effects	*Luisia zeylanica*
Wound healing and tissue generation	*Luisia teres, Luisia tristis*
Antidepressant and anxiolytic properties	*Luisia viridiflora, Luisia pulchella*
Anticancer potential	*Luisia zeylanica*

1. **Pain Relief and Anti-inflammatory Effects:** *Luisia* species, such as *Luisia brachystachys* and *Luisia teres*, have demonstrated analgesic and anti-inflammatory properties. Traditional Ayurvedic medicine has utilized these orchids for centuries to alleviate pain and inflammation associated with various ailments. Studies have shown that extracts from *Luisia* orchids can inhibit the production of inflammatory mediators, reducing pain and inflammation.
2. **Antibacterial and Antimicrobial Properties:** *Luisia* orchids have exhibited antibacterial and antimicrobial activities against a wide spectrum of microorganisms. Extracts from these orchids have been found to possess potent antimicrobial compounds that can combat bacterial and fungal infections. *Luisia* species, like *Luisia javanica* and *Luisia tristis*, have shown promise as potential natural antibiotics, offering an alternative approach to combatting drug-resistant microbes.
3. **Antioxidant and Neuroprotective Effects:** *Luisia* orchids contain a wealth of antioxidant compounds, including flavonoids and phenolic acids. These compounds can neutralize free radicals, protect cells from oxidative damage, and reduce the risk of chronic diseases. Additionally, studies have demonstrated the neuroprotective effects of *Luisia* extracts, suggesting their potential in combating neurodegenerative disorders such as Alzheimer's and Parkinson's diseases.
4. **Antidepressant and Anxiolytic Properties:** Research has revealed the antidepressant and anxiolytic effects of *Luisia* orchid extracts. Compounds isolated from these orchids have shown promise in reducing symptoms of depression and anxiety. *Luisia* species, such as *Luisia viridiflora* and *Luisia pulchella*, have been traditionally used in folk medicine to alleviate stress, anxiety, and mood disorders.
5. **Wound Healing and Tissue Regeneration:** *Luisia* orchids possess remarkable wound healing and tissue regeneration properties. Traditional Chinese medicine has long employed *Luisia* extracts to promote wound healing and reduce scarring. Studies have confirmed the efficacy of *Luisia* extracts in accelerating wound closure, enhancing tissue regeneration, and reducing inflammation.
6. **Anticancer Potential:** Some *Luisia* species have exhibited anticancer properties. Research has shown that extracts from these orchids can inhibit the growth and proliferation of cancer cells. *Luisia* extracts have demonstrated anti-tumor activity against various cancer cell lines, suggesting their potential as adjuvants in cancer therapy.

11.7 CURRENT CONTEXT AND FUTURE

The growing interest in exploring and validating their therapeutic potential through scientific research *Luisia* orchids are slowly and steadily making their place in the herbal drug industry and research.

Traditionally, *Luisia* orchids have been used in various cultures for their purported medicinal properties, but modern science is now providing evidence to support these claims and uncover new applications.

Research has demonstrated the presence of bioactive compounds in *Luisia* orchids, including alkaloids, flavonoids, and terpenoids, which are known for their diverse pharmacological activities. These compounds have shown promise in various areas. Studies have found that extracts from *Luisia* orchids possess anti-inflammatory and analgesic properties, suggesting their potential to alleviate pain and inflammation associated with conditions like arthritis and chronic pain. Antimicrobial and antifungal properties against a range of pathogens have been observed in *Luisia* species making them potential candidates for developing natural antibiotics and antifungal agents to combat resistant microbes. *Luisia* orchids contain a wealth of antioxidant compounds that protect cells from oxidative damage, accounting for their potential anticancer property. Additionally, studies have shown that *Luisia* extracts may have neuroprotective effects, potentially beneficial in neurodegenerative disorders like Alzheimer's and Parkinson's diseases. Researches also suggest that *Luisia* orchid extracts may have antidepressant and anxiolytic effects, offering a natural approach to managing stress, anxiety, and mood disorders. Traditional medicine has long used *Luisia* extracts to promote wound healing and reduce scarring. This has been validated by modern studies that have confirmed the efficacy of *Luisia* extracts in accelerating wound closure and enhancing tissue regeneration.

Despite these promising findings, further research is necessary to fully understand the mechanisms of action, safety, and efficacy of *Luisia* orchids in various medicinal applications. Clinical trials are needed to evaluate their effectiveness and safety in humans before they can be widely used in clinical practice.

The current context of *Luisia* orchids' medicinal uses is one of exploration, validation, and cautious optimism. As research continues to delve into the therapeutic potential of these orchids, there is hope that they may provide valuable contributions to modern medicine, offering natural remedies for a range of ailments and promoting overall well-being.

11.8 CONCLUSION

The genus *Luisia*, with its diverse species, holds immense potential in the field of medicine and healthcare. From pain relief and anti-inflammatory effects to antibacterial, antioxidant, and neuroprotective properties, *Luisia* orchids have demonstrated a wide range of therapeutic applications. Their potential in combating chronic diseases, promoting wound healing, and offering hope in the fight against cancer is truly remarkable. While traditional medicine has long recognized the curative and restorative powers of *Luisia* orchids, modern scientific research is now validating and expanding our understanding of their medicinal value. As research continues to unravel the secrets held within these delicate blooms, the future of *Luisia* orchids as a source of natural remedies and pharmaceuticals looks promising. However, it is crucial to approach the use of *Luisia* orchids with caution and seek guidance from qualified healthcare professionals, as some species may have potential toxicities or contraindications. Further exploration and responsible utilization of these orchids can lead to the development of novel treatments and therapies, contributing to the well-being of humanity and honoring the ancient wisdom that has guided their use for centuries.

REFERENCES

Australian National Botanic Garden (ANBG). www.anbg.gov.au/cpbr/cd-keys/RFKOrchids/key/rfkorchids/Media/Html/genera/Luisia.htm#:~:text=Luisia&text=Erect%20to%20straggly%20epiphytes%20with,usually%20thickened%20and%20club%20%2Dshaped.

Balachandar, M., Ravi, R. K., Ranjithamani, A., & Muthukumar, T. 2019. Comparative vegetative anatomy and mycorrhizal morphology of three South Indian *Luisia* species (Orchidaceae) with the note on their epiphytic adaptations. *Flora*, 251, pp. 39–61.

De, L. C. 2022. Chemical ingredients of important 75 medicinal orchids. *IJCS*, 6(1), pp. 73–79.
Gutiérrez, R. M. P. 2010. Orchids: A review of uses in traditional medicine, its phytochemistry and pharmacology. *Journal of Medicinal Plants Research*, 4(8), pp. 592–638.
Hossain, M. M. 2009. Traditional therapeutic uses of some indigenous orchids of Bangladesh. *Medicinal and Aromatic Plant Science and Biotechnology*, 42(1), pp. 101–106.
Jhansi, K., Rahamtulla, M., & Khasim, S. 2023. Phytochemicals and pharmacological potential of *Acampe ochracea* and *A. praemorsa* (Orchidaceae): An overview. *Phytochemical Composition and Pharmacy of Medicinal Plants*, 2, p. 189.
Khasim, S. M., Jhansi, K., & Venkatesh, R. 2019. Anticancer efficacy of *Luisia zeylanica* Lindl., an epiphytic orchid from Eastern Ghats of Andhra Pradesh, India-an in vitro approach. *Acta Horticulturae*, 1298, pp. 371–376. https://doi.org/10.17660/ActaHortic.2020.1298.51
Ma, L., Zhou, C. Y., Chen, J. L., Liu, D. K., Lan, S., & Liu, Z. J. 2023. Comparative analysis of *Luisia* (Aeridinae, Orchidaceae) plastomes shed light on plastomes evolution and barcodes investigation. *Genes*, 15(1), p. 20.
Plants World Online. https://powo.science.kew.org/taxon/urn:lsid:ipni.org:names:641852-1#:~:text=The%20native%20range%20of%20this,in%20the%20wet%20tropical%20biome.
Rahman, M., & Huda, M. K. 2021. Exploration of phytochemical, antioxidant and anti-inflammatory efficacy of the ethnomedicinal uses of ten orchids of Bangladesh. *Advancement in Medicinal Plant Research*, 9(2), pp. 30–39.
Rajeshbhai, V. A. I. 2022. A review article on: An overview of medicinal importance of orchids. *Acta Scientific Medical Sciences*, 28, p. 36.
Saravanan, K., Prabha, L., Chidambaram, K., Subramanian, A., Kalirajan, A., Veeraiyan, N., Seventhilingam, K., Karthik, S., Abdhul, K., & Ramalakshmi, C. 2023. Exploring the anti-bacterial potential of orchid-derived silver nanoparticles. *Nano Biomedicine & Engineering*, 15(4), pp. 416–424.
Sethuraman, S. P., & Ramachandran, K. P. 2022. Pharmacognostical standardisation of an epiphytic orchid, *Luisia tenuifolia* Blume. *Journal of Pharmacy & Pharmacognosy Research*, 10(1), pp. 113–127.
Singh, A., & Duggal, S. 2009. Medicinal orchids - An overview. *Ethnobotanical Leaflets*, 2009(3), p. 3.
Sohag, S. I., Hoque, M. M., & Huda, M. K. 2017. Phytochemical screening and antioxidant activity of rare medicinal orchid *Luisia zeylanica* Lindl. *Journal of Pharmacognosy and Phytochemistry*, 6(4), pp. 688–692.
Teoh, E. S. 2016. *Medicinal Orchids of Asia*. Cham: Springer.

12 The Significance and Utilization of *Pholidota* Orchids in Medicinal Preparations

Mukti Ram Poudeyal, Khageshwari Saud, Mukti Ram Paudel, and Uday Kumar Singh

12.1 INTRODUCTION

Orchids are widely utilized in traditional medicinal procedures throughout ethnic communities due to their reputation for healing qualities. At the same time, orchids have also become a significant part of the lives of people in the rural world because of their visually striking and fascinating flower shape and size, longevity, and increasing market prices. Ecologically, orchids are low-multiplying and slow-growing plants, need highly specific mycorrhizal association for germination, and need specific pollinators for pollination. Thus, they are highly sensitive creatures of nature, and their charismatic presence signifies a vibrant and lively ecosystem. A variety of phytochemicals, including coumarins, amino acids, sugars, starch, glucoside, terpenoids, stilbenoids, flavonoids, saponins, tannins, steroids and mucilage possess a wide range of biological activities that have been receiving tremendous attention in therapeutics (Vasas 2021). Owing to their natural beauty and ecological and medicinal values, numerous orchid species are threatened by unsustainable (often illegal and undocumented) harvesting for horticulture, food, or medicine (Pant and Raskoti 2013).

Pholidota (Family: Orchidaceae, Sub-family: Epidendroideaee) is one of the popular often epiphytic (rarely saxicolous) taxa among the orchid species in Nepal (440 orchid species belonging to 95 genera) (Shrestha et al. 2022). Epithetically, the generic name is derived from the Greek word '*Pholidotos*', which means imbricate bracts of the inflorescence. Due to high morphological intra-generic diversity and arbitrarily placement of the species, genus *Pholidota* is one of the taxonomically disputed orchids which makes generalization difficult, and can be considered as the best research material for the taxonomic revision in coming days ahead (Chase et al. 2021; Li et al. 2019). Morphologically, most of the species are allied with *Coelogyne* (subtribe: *Coelogyninae* and tribe: Arethuseae), and thus, some literature merged whole or partial number of species of *Pholidota* into *Coelogyne* (for instance, Shrestha et al. 2022; Chase et al. 2021). However, saccate hypochile has long been utilized to distinguish *Pholidota* as a diagnostic character. Because of its stem-like jointed pseudobulbs, it is also known as articulated orchid, jointed orchid, rattlesnake, and necklace orchid. Moreover, molecular systematics (plastid genomic data) of the *Pholidota* stated that species of the genus are polyphyletic, overlap not only with *Coelogyne*, but also with *Bulleyia* and *Panisea* (Li et al. 2019).

De Vogel (1988) described at least 29 species of *Pholidota* in different parts of the world, and he mentioned almost 15 species with doubtful status. Based on information provided by de Vogel (1988), Liu et al. (2002), Li et al. (2019), and Dang et al. (2021), there are 33 species (Table 12.1) found to be recorded with taxonomic validation. The species are mainly distributed in the Himalayas and from southeast China to Queensland. However, only six species are reported from Nepal Himalaya and its territory: *Pholidota articulata* (Figure 12.1a), *Pholidota griffithii*, *Pholidota imbricata*, *Pholidota pallida* (Figure 12.1b), *Pholidota protracta*, and *Pholidota recurva*

DOI: 10.1201/b23375-12

TABLE 12.1
Global Distribution Range of *Pholidota* Species

S. No.	Species	Distribution
1	*Pholidota advena* (Par. and Rchb. f.) Hooker f.	Thailand, Malaysia, Philippines, Indonesia, Myanmar, and China
2	*Pholidota aidiolepis* Seidenf. and de Vogel	N. Myanmar, Peninsula Thailand to Peninsula Malaysia.
3	*Pholidota articulata* Lindl.	Indo-China, Peninsular Malaysia
4	*Pholidota carnea* (Blume) Lindl.	Peninsula Thailand to New Guinea
5	*Pholidota chinensis* Lindl.	Sikkim to S. China and Indo-China
6	*Pholidota camelostalix* Rchb.f.	Sumatera to Lesser Sunda Islands
7	*Pholidota convallariae* (C.S.P.Parish and Rchb.f.) Hook.f.	E. Himalaya to China (SW. Yunnan) and W. Malesia
8	*Pholidota cantonensis* Rolfe	Endemic to southern China
9	*Pholidota clemensii* Ames	Borneo
10	*Pholidota gibbosa* (Blume) de Vriese	Malesia to Solomon Islands
11	*Pholidota globosa* (Blume) Lindley	Sumatera to Lesser Sunda Islands (Bali)
12	*Pholidota guibertiae* Finet	S. Vietnam
13	*Pholidota imbricata* Hook	Tropical & Subtropical Asia to SW. Pacific
14	*Pholidota leveilleana* Schlechter	Endemic to southwestern China
15	*Pholidota longibulba* Holttum	SW. Thailand to Peninsula Malaysia
16	*Pholidota longilabra* de Vogel	N. Sumatera
17	*Pholidota longipes* S.C. Chen and Z.H. Tsi.	China (SE. Yunnan)
18	*Pholidota mediocris* de Vogel	Borneo
19	*Pholidota missionariorum* Gagnep.	Bhutan to S. Central China and N. Indo-China
20	*Pholidota nervosa* (Blume) Rchb. f.	Sumatera to W. Jawa
21	*Pholidota niana* Y.T.Liu, R.Li, and C.L. Long	China (SE. Yunnan)
22	*Pholidota roseans* Schltr.	China (S. Guizhou) to N. Vietnam
23	*Pholidota pallida* Lindl.	Himalaya to S. Central China and Indo-China
24	*Pholidota pectinata* Ames	Borneo
25	*Pholidota protracta* Hook. f.	Nepal to China (NW. Yunnan) and N. Myanmar
26	*Pholidota recurva* Lindl.	Nepal to S. Central China and Indo-China
27	*Pholidota rubra* Lindley	Sikkim to Indo-China
28	*Pholidota schweinfurthiana L.O. Williams*	Borneo
29	*Pholidota subcalceata* Gagnepain	Endemic to southern Vietnam
30	*Pholidota sulcata* J.J. Smith	Borneo
31	*Pholidota ventricosa* (Bl.) Rchb.f.	Vietnam, Malesia to New Guinea
32	*Pholidota wenshanica* S.C. Chen and Z.H. Tsi	Endemic to southwestern China
33	*Pholidota yunnanensis* Rolfe	S. China to N. Indo-China

(Shrestha et al. 2022; Rokaya et al. 2013). Their morphological complexities on pseudobulbs possess moisture retention and water reservation potential, exhibiting a higher level of ecological adaptation. For ages, people in the Himalayas have utilized folk herbal medicine made from plants of this genus, which has a wide range of medical benefits. It is widely employed for health complaints

FIGURE 12.1 The genus *Pholidota*. *P. articulata* (a) and *P. pallida* (b).

Source: Photo Credit: Bijaya Pant and Mukti Ram Paudel

ranging from heat-clearing to detoxifying, treating tumors, anxiolytic activities, sedative, hypnotic, anticonvulsive, and neuroprotective. The plant includes a wealth of phytochemicals that have antibacterial, anti-inflammatory, and antioxidant properties. These substances might potentially play a major role in developing medicines in the future. From the ethyl acetate extract of *Pholidota*, a well-known functional food and traditional Chinese medicine, phenoldonone, a stilbene credible derivative with strong anticancer activity (Table 12.2), is extracted (Chen et al. 2024).

TABLE 12.2
Biological Activities of Species of *Pholidota* Orchids and Their Potential for Drug Development

S. No.	Species	Biological Activities	References
1	*Pholidota articulata* Lindl.	• **Antioxidant** [IC_{50} of methanolic extract (196.03 μg/ml) for DPPH scavenge, Petroleum ether extract scavenge 21.41% of DPPH] • **Antibacterial** [MIC of methanol extract against *Salmonella enterica* (50 μg/ml); Ethyl acetate extract against *Klebsiella pneumonia* (18 mm), *S. enterica* (16 mm), *E. coli* (14 mm)] • **Antifungal** [crude extract and methanolic extract against *T. ruberum*, *A. flavus*, and *A. niger*, ethyl acetate and petroleum ether extracts against *C. albicans* and *P. glaucum*] • Antidiabetic activity [IC50 of methanol extract of leaves (1268.80 μg/ml), pseudobulb (1539.61 μg/ml) for α-amylase inhibition] • Osteogenic activity [ethanol extract and Oxoflavidin]	Debnath and Kumaria 2023; Joshi et al. 2023; Kaushik 2019; Singh et al. 2016a, 2016b, 2018; Sharma et al. 2015
2	*Pholidota chinensis* Lindl.	• **Antioxidant** [IC_{50} of ethyl acetate extract for DPPH (1.43 mM AAE/g); for (ABTS 5.38 mM TE/g); for FRAP (2.58 mM FSHE/g); EC_{50} of polysaccharide for superoxide (1.09 mg/ml) and for hydroxyl (0.11 mg/ml)] • **Antidiabetic** [IC_{50} of ethyl acetate extract for α-glucosidase inhibition (13.29 mM ACAE/g)] • **Anti-inflammatory** [IC_{50} of ethyl acetate extract (18.71 μg/ml); IC_{50} of pholidotol A and pholidotol B of Nitric oxide inhibition (24.3 and 17.1 μM); IC_{50} of five dihydrophenanthropyrans (0.23 to 7.63 μM)] • **Analgesic** • **Anti-fatigue** • **Anti-hypoxia** • **Anticancer** [IC_{50} of two phenanthrenes on MCF-7 cells (8.1 and 7.5 μM); IC_{50} of polysaccharide on human colon cancer cells (69.54 μg/ml)] • **GABAA receptor modulators** [EC_{50} of dihydrostilbene (52.51 μM)] • **Antibacterial** [MIC of 2-methoxy-3,7-dihydroxy-5H phenanthro[4,5-bcd] pyran (68.39 μM) against *Staphylococcus aureus*]	Chen et al. 2024; Ti et al. 2022; Wang et al. 2006, 2007, 2021; Ren et al. 2020; Luo et al. 2018; Yang et al. 2016; Rueda et al. 2014; Wu et al. 2008

(Continued)

TABLE 12.2 *(Continued)*
Biological Activities of Species of *Pholidota* Orchids and Their Potential for Drug Development

S. No.	Species	Biological Activities	References
3	*Pholidota cantonensis* Rolfe	• **Antioxidant** • **Anti-inflammatory** • **Antitumor** • **Antibacterial** • **NO inhibition** • **Anti-Alzheimer's** • **Cytotoxicity** [IC_{50} of pholidonone on human gastric cancer cells (17.28 µg/ml), AGS cells (25.20 µg/ml); IC_{50} of phocantone, phocantoside A, tanshinone IIA, and syringate on mouse leukemia cancer cells (13.37 to 27.50 µg/ml)]	Liu et al. 2019, 2021; Li et al. 2017
4	*Pholidota imbricata* Hook	• **Antibacterial** [MIC and MBC values of crude extracts against *V. cholera* were 62.5 mg/ml and 125 mg/ml, respectively] • **Antifungal** [*Candida albicans*, *Rhyzopus stolonifera*, *Muco*r spp.] • **NO inhibition** • **Antioxidant**	Wang et al. 2014; Marasini and Joshi 2012
5	*Pholidota leveilleana* Schlechter	Cytotoxic and free radical scavenging activity (9,10-dihydrophenanthropyrans, and stilbenes and bibenzyls)	Yuan et al. 2013
6	*Pholidota pallida* Lindl.	• **Antioxidant** [98.94% at 50 µg/ml of butanol fraction] • **Anti-inflammatory** [79.16% by dichloromethane extract] • **Antibacterial** [*Staphylococcus aureus* (dichloromethane fraction of root; 17.5 mm), *Bacillus subtilis* (dichloromethane fraction of bulb; 16.5 mm), *Salmonella typhi* (butanol fraction of root; 12 mm), *B. cereus* (butanol fraction of root; 9.5 mm); MIC ranging from 0.312 to 2.5 mg/ml against *Escherichia coli*, *Salmonella typhi*, *Bacillus subtilis*, *Shigella flexneri*, and *Pseudomonas aeruginosa*] • **Antifungal** [Leaf nanoparticles showed inhibition against *Fusarium* sp.]	Madhavamurthy et al. 2022; Sharone and Chithra 2021; Akter et al. 2019; Hoque et al. 2016; Nagananda et al. 2014
7	*Pholidota protracta* Hook. f.	• Scavenging free radicals and cytotoxic substances (9,10-dihydrophenanthropyrans, 9,10-dihydrophenanthrenes)	Zhan et al. 2010
8	*Pholidota yunnanensis* Rolfe	• **Antioxidant** [EC50 of phenanthrenes (8.8 to 55.9 µM)] • **NO inhibition** [IC_{50} of eight stilbenoids, three dihydrophenanthrenes, three stilbenes, and one bibenzyl derivative showed inhibitory effects on NO production without cytotoxicity in a murine macrophage-like cell line (1.5 to 7.77 µM)] • **Cytotoxicity** [phoyunbene B inhibit the growth of hepatocellular carcinoma cells (HepG2); phoyunnanin C showed cytotoxic effect]	Qi et al. 2024; Dong et al. 2013; Wang et al. 2012; Guo et al. 2006, 2007

Source: Based on de Vogel 1988; Liu et al. 2002; Li et al. 2019; Dang et al. 2021

The considerable diversity of habitat types and varied climates that result from the wide altitudinal difference (60–8,848 m above the sea level) in Nepal enables the environment to support a large variety of orchids with varying chemical compositions, which provides enormous opportunities for innovation. Advances in biotechnological techniques and phytochemical analysis have significantly improved the production of bioactive secondary metabolites from different *Pholidota* species. However, in the context of Nepal, there is a dearth of information, often sporadic and limitedly documented, on relevant phytochemical and pharmacological studies on this orchid, which is highly demanding for medicinal preparations, which limits its use and advancement. The current study aimed to provide a theoretical foundation for the development and application of the medicinal preparation of this plant.

12.2 ETHNOBOTANICAL EXPLORATION OF *PHOLIDOTA*

In various regions of the world, along with medicinal and ornamental use, orchids have been utilized as food, flavoring additives, and herbal tea (Manandhar 2002; Pal et al. 2022). The number of ethnic groups in Nepal that use orchids as medicine is unclear, although it's thought that practically all of them do (25 major ethnic groups plus a few minor tribal tribes), at least in some form. Oral forms of orchid consumption include powder, decoction, tincture, and juice. They can be consumed on their own or combined with other foods, such as honey, to make them more enticing.

A mixture of ground paste of the whole *Pholidota articulata* plant and soaked rice is used to cure joint fractures and dislocations. Also, its pseudobulbs contain a thick gum-like substance that can be extracted for the tonic, and it has been replaced with *Jivanti* (*Leptadenia reticulata*) in Ayurveda (Debnath and Kumaria 2023). Plant paste of *Pholidota imbricata* is consumed to lower the fever and rheumatic pain, and powder is administered as a tonic and applied externally to heal fractures (Pal et al. 2022). Applying coarsely macerated pseudobulbs in mustard oil to joints can alleviate rheumatic pain and boils (Manandhar 2002; Pant and Raskoti 2013). Its crushed pseudobulbs are used for spine ejection (Vaidya 2019). The pseudobulb extract is used to cure abdominal pain and rheumatism. *Pholidota pallida* plant paste is used to lower fever, the powder is used to treat sleep apnea and to lessen stomachache, the juice is used to alleviate pain in the navel, roots are used to treat rheumatism, and the pseudobulb is used to treat intestinal worms (Vaidya 2019). Roots of *P. pallida* are utilized for rheumatism, while pseudobulbs are used to treat intestinal worms and pain in the stomach (Pal et al. 2022). The whole plant of *P. yunnanensis* is reported to be used to treat cough, external injuries, rheumatism, stomachache, and trauma (Bandi and Lee 2011). Pseudobulb of *Pholidota chinensis* is sedative and anticonvulsant (Bandi and Lee 2011) and also used in the treatment of high blood pressure, scrofula, toothache, headache, asthmatic cough, tuberculosis, and dysentery, and rhizome paste is used in finger abscess (Teoh 2019; Pal et al. 2022). Its juice mixed in the coconut kernel is applied to inflamed skin.

12.3 CHEMISTRY IN *PHOLIDOTA*

Only from orchids have prenylated compounds and stilbene/dihydrostilbene-connected phenanthrenes been found, particularly from the species *Pholidota* (further elaborated in Table 12.2). Three stilbenoids and 15 stilbene derivatives, dihydrophenanthrene derivatives, were isolated from *Pholidota yunnanensis* (Qi et al. 2024; Dong et al. 2013; Guo et al. 2006, 2007). Phenolic compounds, stilbene derivatives (such as pholidotol A and pholidotol B), phenanthrene derivatives, dihydrophenanthropyran derivatives, stilbenoids (coelonin, batatasin III, and pholidotol D) dihydrostilbenes, and polysaccharides were isolated from *P. chinensis* (Chen et al. 2024; Ti et al. 2022; Wang et al. 2006, 2007, 2021; Ren et al. 2020; Luo et al. 2018; Yang et al. 2016; Rueda et al. 2014; Wu et al. 2008). Secondary metabolites like coelogin, flaccidin, 7-hydroxy-2,3,4-trimethoxy-9,10-dihydrophenanthrene, 3-hydroxy-5-methoxybibenzyl, dihydroconiferyl alcohol, 3-hydroxy benzenemethanol, hydroquinone, and salicylic acid were isolated from *P. protracta* (Zhan et al. 2010). Furthermore, flavidin, oxoflavidin, phenenthrenes, and bibenzyls were isolated from *P. articulata* (Zhu et al. 2022; Sharma et al. 2015; Majumder et al. 1982). Fourteen chemicals were extracted from the *P. leveilleana*: pinosylvin monomethyl ether, thunalbene, methyl vanillate, p-hydroxybenzaldehyde, vanillin, p-hydroxybenzoic acid, syringaldehyde, coelonin, 9,10-dihydro-2,4-dihydroxy-7-methoxyphenanthrene, lusianthridin, flavanthrin, and (—)-pinoresinol (Yuan et al. 2013). Similarly, phenanthrene derivatives phoimbrtol A, loddigesiinol B, shanciol B, medioresinol, pinoresinol, quercetin 3-O-L-arabinofuranoside, luteolin 7-O-α-glucoside, and platycaryanin D were isolated from *P. imbricata* (Wang et al. 2014). Stilbene (pholidonone), dihydrophenanthrenes (phocantol and phocantone), diterpenoid glycoside (phocantoside A and phocantoside B), and bibenzyl derivatives were isolated from *P. cantonensis* (Liu et al. 2019, 2021; Li et al. 2017). Similarly, coelonin, lusianthridin, flavanthrin, batatasin, bibenzyl derivative, gigantol, hydroxytyrosyl butyrate, taraxerone, and friedelin, including phenanthrenes, steroid, and terpenoids, were isolated from *Pholidota pallida* (Yu et al. 2021). Imbricatin, Coelonin, Pholidotin (=cycloart-25-en-yl (E)-p-hydroxycinnamate), and 24-Methylidenecycloartan-3-yl (E)-p-hydroxycinnamate were isolated from *P. rubra* (Bandi and Lee 2011).

12.4 BIOLOGICAL ACTIVITIES OF *PHOLIDOTA*

Currently, 33 species have been recognized in the genus *Pholidota*; out of them, eight species are of great medicinal importance to our findings (Table 12.2). *P. articulata* has potent antioxidant, antibacterial, antifungal, antidiabetic, and osteogenic activities (Debnath and Kumaria 2023; Joshi et al. 2023; Kaushik 2019; Singh et al. 2016a, 2016b, 2018; Sharma et al. 2015). *P. chinensis* has antioxidant, antidiabetic, anti-inflammatory, analgesic, antibacterial, anticancer, anti-fatigue, and anti-hypoxia activities (Chen et al. 2024; Ti et al. 2022; Wang et al. 2006, 2007, 2021; Ren et al. 2020; Luo et al. 2018; Yang et al. 2016; Rueda et al. 2014; Wu et al. 2008). *P. cantonensis* has antioxidant, anti-inflammatory, antitumor, antibacterial, NO inhibition, anti-Alzheimer's, and cytotoxicity activities (Liu et al. 2019, 2021; Li et al. 2017). *P. imbricata* has antioxidant, antifungal, antibacterial, and NO inhibition activities (Wang et al. 2014; Marasini and Joshi 2012). *P. pallida* has antioxidant, antibacterial, antifungal, and anti-inflammatory activities (Madhavamurthy et al. 2022; Sharone and Chithra 2021; Akter et al. 2019; Hoque et al. 2016; Nagananda et al. 2014). *P. yunnanensis* has antioxidant, cytotoxicity, and NO inhibition activities (Qi et al. 2024; Dong et al. 2013; Wang et al. 2012; Guo et al. 2006, 2007).

12.5 CONCLUSION

The combined findings imply that species of this epiphytic orchid *Pholidota* have multidimensional value including food additives, flavoring, and ornamental uses. A multitude of phytochemicals (significantly such as stilbenoids, benzoxepins, triterpenes, steroids, phenolic compounds, and lignans) in extracts or pure compounds from different parts of the plant confer anti-inflammatory, antioxidant, cytotoxicity, antifungal, antitumor, and antibacterial properties. These phytochemicals could eventually turn out to be a major source of value-added pharmacological, food, and cosmetic products for industrial scale; plethora of anticancer drugs; and relevant medications in the future.

REFERENCES

Akter, S., Huda, M. K., Rahman, M., Hoque, M. M., & Eva, T. A. (2019). Phytochemical profiling and bioactivities of *Pholidota pallida* Lindl. *International Journal of Advanced Research in Botany*, *5*, 8–13.

Bandi, A. K. R., & Lee, D.-U. (2011). Chemical constituents and bioactivities of plants from the genus *Pholidota*. *Chemistry and Biodiversity*, *8*(8), 1400–1409.

Chase, M. W., Gravendeel, B., Sulistyo, B. P., Wati, R., & Schuiteman, A. (2021). Expansion of the orchid genus *Coelogyne* (Arethuseae; Epidendroideae) to include *Bracisepalum, Bulleyia, Chelonistele, Dendrochilum, Dickasonia, Entomophobia, Geesinkorchis, Gynoglottis, Ischnogyne, Nabaluia, Neogyna, Otochilus, Panisea* and *Pholidota. Phytotaxa, 510*(2), 94–134.

Chen, Y. C., Pan, W. X., Wang, Y. H., Tsai, C. M., Hwang, T. L., & Lam, S. H. (2024). Dihydrophenanthropyrans derived from the pseudobulbs of *Pholidota chinensis* alleviates neutrophilic inflammation by inhibiting MAPKs and calcium. *Fitoterapia, 176*, 106015.

Dang, M. Q., Ormerod, P., Bui, V. H., Truong, Q. T., Duong, M. T. T. N., Dang, V. S., & Truong, B. V. (2021). *Pholidota longipes* SC Chen & ZH Tsi, a new addition for flora of Vietnam. *Feddes Repertorium, 132*(3), 228–231.

de Vogel, E. F. (1988). Revisions in Coelogyninae (Orchidaceae) III. The genus *Pholidota. Orchid Monograph, 3*, 1–116.

Debnath, S., & Kumaria, S. (2023). Insights into the phytochemical accumulation, antioxidant potential and genetic stability in the in vitro regenerants of *Pholidota articulata* Lindl., an endangered orchid of medicinal importance. *South African Journal of Botany, 152*, 313–320.

Dong, F. W., Fan, W. W., Xu, F. Q., Wan, Q. L., Su, J., Li, Y., Zhou, L., Zhou, J., & Hu, J. M. (2013). Inhibitory activities on nitric oxide production of stilbenoids from *Pholidota yunnanensis. Journal of Asian Natural Products Research, 15*(12), 1256–1264.

Guo, X. Y., Wang, J., Wang, N. L., Kitanaka, S., Liu, H. W., & Yao, X. S. (2006). New stilbenoids from *Pholidota yunnanensis* and their inhibitory effects on nitric oxide production. *Chemical and Pharmaceutical Bulletin, 54*, 21–25.

Guo, X. Y., Wang, J., Wang, N. L., Kitanaka, S., & Yao, X. S. (2007). 9,10-dihydrophenanthrene derivatives from *Pholidota yunnanensis* and scavenging activity on DPPH free radical. *Journal of Asian Natural Products Research, 9*(2), 165–174.

Hoque, M. M., Khaleda, L., & Al-Forkan, M. (2016). Evaluation of pharmaceutical properties on microbial activities of some important medicinal orchids of Bangladesh. *Journal of Pharmacognosy and Phytochemistry, 5*(2), 265–269.

Joshi, P. R., Joshi, S., Paudel, M. R., & Pant, B. (2023). Screening of in vitro α-amylase inhibitory activity of wild orchids of Nepal. *Journal of Nepal Biotechnology Association, 4*(1), 1–7.

Kaushik, P. (2019). Antibacterial potential of the Himalayan orchids. *Journal of Orchid Society of India, 33*, 11–22.

Li, B., Ali, Z., Chan, M., Li, J., Wang, M., Abe, N., Wu, C. R., Khan, I. A., Wang, W., & Li, S. X. (2017). Chemical constituents of *Pholidota cantonensis. Phytochemistry, 137*, 132–138.

Li, L., Qin, M., Wang, W. Y., Zeng, S. J., Zhang, G. Q., & Liu, Z. J. (2019). The taxonomic identities of *Pholidota wenshanica* and *P. subcalceata* (Orchidaceae, Coelogyninae). *PhytoKeys, 136*, 97–106.

Liu, L., Wang, W., Zhao, Z., Hu, C., Tao, L., & Zhang, X. (2019). Pholidonone, an active stilbene derivative from *Pholidota cantonensis*, exhibits pro-apoptotic effect via induction of endoplasmic reticulum stress in human gastric cancer. *Food & Nutrition Research, 63*, 3553 (http://dx.doi.org/10.29219/fnr.v63.3553).

Liu, L., Zou, M., Zeng, K., Ye, X., Wang, R., Wang, W., & Zhang, X. (2021). Chemical constituents and their antioxidant, anti-Inflammatory and anti-acetylcholinesterase activities from *Pholidota cantonensis. Plant Foods for Human Nutrition, 76*, 105–110.

Liu, Y. T., Li, R., & Long, C. L. (2002). *Pholidota niana* (Orchidaceae), a news species from southeastern Yunnan, China. *Annales Botanici Fennici, 39*, 227–229.

Luo, D., Wang, Z., Li, Z., & Yu, X. Q. (2018). Structure of an entangled heteropolysaccharide from *Pholidota chinensis* Lindl and its antioxidant and anti-cancer properties. *International Journal of Biological Macromolecules, 112*, 921–928.

Madhavamurthy, H., Chikkamadaiah, M., & Suryanarayana, S. M. (2022). Screening of biological activity of selected medicinal orchids of Western Ghats, Karnataka, India. *INNOSC Theranostics and Pharmacological Sciences, 5*(2), 22–31.

Majumder, P. L., Datta, N., Sarkar, A. K., & Chakraborti, J. (1982). Flavidin. A novel 9, 10-dihydrophenanthrene derivative of the orchids *Coelogyne flavida, Pholidota articulata* and *Otochilus fusca. Journal of Natural Products, 45*(6), 730–732.

Manandhar, N. P. (2002). *Plants and People of Nepal.* Portland: Timber Press.

Marasini, R., & Joshi, S. (2012). Antibacterial and antifungal activity of medicinal orchids growing in Nepal. *Journal of Nepal Chemical Society, 29*, 104–109.

Nagananda, G. S., Patil, A., Kambli, J. V., & Rajath, S. (2014). Phytochemical evaluation and in vitro free radical scavenging activity of cold and hot successive pseudobulb extracts of medicinally important orchid *Pholidota pallida* Lindl. *Advances in Bioresearch*, *5*(1), 100–105.

Pal, R., Meena, N. K., Dayamma, M., & Singh, D. R. (2022). Ethnobotany and recent advances in Indian medicinal orchids. In: Mérillon, J. M., & Kodja, H. (eds.), *Orchids Phytochemistry, Biology and Horticulture*. Cham: Springer, pp. 361–387.

Pant, B., & Raskoti, B. B. (2013). *Medicinal Orchids of Nepal*. Kathmandu, Nepal: Himalayan Map House.

Qi, J., Jia, Y., Zhou, D., Chen, G., Hao, J., Mi, Y., Xu, L., Lin, B., Hou, Y., & Li, N. (2024). Axially chiral dihydrophenanthrene dimers from *Pholidota yunnanensis* with anti-neuroinflammatory activities. *Bioorganic Chemistry*, *150*, 107570.

Ren, M., Xu, W., Zhang, Y., Ni, L., Lin, Y., Zhang, X., & Huang, M. (2020). Qualitative and quantitative analysis of phenolic compounds by UPLC-MS/MS and biological activities of *Pholidota chinensis* Lindl. *Journal of Pharmaceutical and Biomedical Analysis*, *187*, 113350.

Rokaya, M. B., Raskoti, B. B., Timsina, B., & Münzbergová, Z. (2013). An annotated checklist of the orchids of Nepal. *Nordic Journal of Botany*, *31*, 511–555.

Rueda, D. C., Schöffmann, A., De Mieri, M., Raith, M., Jähne, E. A., Hering, S., & Hamburger, M. (2014). Identification of dihydrostilbenes in *Pholidota chinensis* as a new scaffold for GABAA receptor modulators. *Bioorganic & Medicinal Chemistry*, *22*(4), 1276–1284.

Sharma, C., Dixit, M., Singh, R., Agrawal, M., Mansoori, M. N., Kureel, J., Singh, D., Narender, T., & Arya, K. R. (2015). Potential osteogenic activity of ethanolic extract and oxoflavidin isolated from *Pholidota articulata* Lindley. *Journal of Ethnopharmacology*, *170*, 57–65.

Sharone, S. G. E., & Chithra, D. B. (2021). Bioactivity of endemic orchids of Western Ghats; *Pholidota pallida* Lindl and *Arundina graminifolia* (D. Don) Hochr. *Plant Archive*, *21*(1), 1214–1220.

Shrestha, K. K., Bhattarai, S., & Bhandari, P. (2022). *Plants of Nepal (Gymnosperms and Angiosperms)*. Kathmandu, Nepal: Heritage Publishers and Distributors Pvt. Ltd.

Singh, D., Sati, S. C., & Sati, M. D. (2016a). Evaluation of free radical scavenging activity of methanolic extract of *Pholidota articulata*. *The Pharma Innovation*, *5*(9, Part A), 6.

Singh, D., Sati, S. C., & Sati, M. D. (2016b). In vitro antimicrobial activity of Himalayan medicinal plant *Pholidota articulata*. *International Journal of Herbal Medicine*, *4*(6), 1–3.

Singh, D., Sati, S. C., & Sati, M. D. (2018). Evaluation of free scavenging activity of petroleum ether extract of *Pholidota articulata* (Orchidaceae). *World Journal of Pharmaceutical Research*, *7*, 1351–1354.

Teoh, E. S. (2019). India: Van Rheede, Caius and others. In: *Orchids as Aphrodisiac, Medicine or Food*. Cham: Springer.

Ti, H., Zhuang, Z., Li, Y., Wei, G., & Wang, F. (2022). Three new phenanthrenes from *Pholidota chinensis* Lindl. and their antibacterial activity. *Natural Product Research*, *36*(8), 2056–2062.

Vaidya, B. N. (2019). Nepal: A global hotspot for medicinal orchids. In: Joshee, N., Dhekney, S., & Parajuli, P. (eds.), *Medicinal Plants*. Cham: Springer.

Vasas, A. (2021). Phenanthrenes from Orchidaceae and their biological activities. In: Merillon, J. M., & Kodja, H. (eds.), *Orchids Phytochemistry, Biology and Horticulture*. Cham: Springer, pp. 1–41.

Wang, D. L., Lin, F. X., Tang, M. M., Zhou, X. M., Yi, J. L., & Wang, T. T. (2021). Two new phenanthrene derivatives from *Pholidota chinensis*. *Phytochemistry Letters*, *43*, 123–125.

Wang, G., Guo, X., Chen, H., Lin, T., Xu, Y., Chen, Q., Liu, J., Zeng, J., Zhang, X. K., & Yao, X. (2012). A resveratrol analog, phoyunbene B, induces G2/M cell cycle arrest and apoptosis in HepG2 liver cancer cells. *Bioorganic & Medicinal Chemistry Letters*, *22*(5), 2114–2118.

Wang, J., Matsuzaki, K., & Kitanaka, S. (2006). Stilbene derivatives from *Pholidota chinensis* and their anti-inflammatory activity. *Chemical and Pharmaceutical Bulletin*, *54*(8), 1216–1218.

Wang, J., Wang, L., & Kitanaka, S. (2007). Stilbene and dihydrophenanthrene derivatives from *Pholidota chinensis* and their nitric oxide inhibitory and radical-scavenging activities. *Journal of Natural Medicines*, *61*, 381–386.

Wang, J., Wang, T., Xie, P., Yin, G., & Li, X. (2014). New phenanthrene derivatives with nitric oxide inhibitory and radical-scavenging activities from *Pholidota imbricata* Hook. *Natural Product Research*, *28*(4), 251–256.

Wu, B., Qu, H., & Cheng, Y. (2008). Cytotoxicity of new stilbenoids from *Pholidota chinensis* and their spin-labeled derivatives. *Chemistry & Biodiversity*, *5*(9), 1803–1810.

Yang, H., Wu, Y., Gan, C., Yue, T., & Yuan, Y. (2016). Characterization and antioxidant activity of a novel polysaccharide from *Pholidota chinensis* Lindl. *Carbohydrate Polymers*, *138*, 327–334.

Yu, J. D., Chen, M. X., Wu, H., Qin, M. L., Zheng, J. R., Li, Y. P., & Zhang, R. P. (2021). Phytochemical and chemotaxonomic study on *Pholidota pallida* lindl. *Biochemical Systematics and Ecology*, *97*, 104290.
Yuan, Z.-P., Zhong, Y.-J., Su, X.-H., Li, C.-Y., Du, H.-J., Wen, Y.-Y., Li, Y.-F., & Liang, B. (2013). Chemical constituents of *Pholidota leveilleana. Biochemical Systematics and Ecology*, *49*, 148–151.
Zhan, R., Min, K., Zhang, Y., Jiang, J., Liu, Y., & Chen, Y. (2010). Phenolic compounds from *Pholidota protracta. Biochemical Systematics and Ecology*, *38*(3), 428.
Zhu, X., Yan, H., Yang, M., Yang, X., & He, L. (2022). Separation of three phenanthrenes and two bibenzyls from chloroform fractions of *Pholidota articulata* Lindl. using HPLC. *Journal of AOAC International*, *105*(1), 282–287.

Index

Note: Page numbers in *italics* indicate a figure and page numbers in **bold** indicate a table on the corresponding page.

D

O

P